Optical MEMS

Optical MEMS

Special Issue Editors

Huikai Xie

Frederic Zamkotsian

MDPI • Basel • Beijing • Wuhan • Barcelona • Belgrade

Special Issue Editors
Huikai Xie
University of Florida
USA

Frederic Zamkotsian
Marseille University
France

Editorial Office
MDPI
St. Alban-Anlage 66
4052 Basel, Switzerland

This is a reprint of articles from the Special Issue published online in the open access journal *Micromachines* (ISSN 2072-666X) from 2018 to 2019 (available at: https://www.mdpi.com/journal/micromachines/special_issues/Optical_MEMS)

For citation purposes, cite each article independently as indicated on the article page online and as indicated below:

LastName, A.A.; LastName, B.B.; LastName, C.C. Article Title. *Journal Name* **Year**, *Article Number*, Page Range.

ISBN 978-3-03921-303-0 (Pbk)
ISBN 978-3-03921-304-7 (PDF)

Cover image courtesy of Liang Zhou and Huikai Xie.

Contents

About the Special Issue Editors

Huikai Xie is a Professor at the Department of Electrical and Computer Engineering of the University of Florida (UF). He received his B.Sc. in microelectronics, M.Sc. in photonics, and Ph.D. in electrical and computer engineering from Beijing Institute of Technology, Tufts University, and Carnegie Mellon University, respectively. Before he joined UF as an assistant professor in 2002, he worked at Tsinghua University (1992–1996), Bosch Corporation (2001), and Akustica Inc. (2002). He has published over 300 technical papers and 11 book chapters and holds 31 US patents. His current research interests include MEMS/NEMS, inertial sensors, microactuators, optical MEMS, optical beam steering, LiDAR, microspectrometers, and optical microendoscopy. He is a fellow of IEEE and SPIE.

Frederic Zamkotsian received the Ph.D. degree in Physics in 1993 from the University of Marseilles (France). Since then, he has worked in the field of optoelectronics and semiconductor physics for optical telecommunication in France and in Japan. In 1998, he joined the Laboratoire d'Astrophysique de Marseille (LAM, Aix-Marseille University, CNRS, CNES), where he is involved in MOEMS-based astronomical instrumentation for ground-based and space telescopes, including conception and characterization of new MOEMS devices, as well as development of new instruments (principal investigator of BATMAN instrument to be placed on 4 m class telescope in 2020 and on 8 m class telescope in 2023). He has published over 200 technical papers and international conference proceedings, as well as 3 book chapters. On MOEMS, his current research interests are in programmable slits for application in multiobject spectroscopy (JWST, European networks, EUCLID, BATMAN), deformable mirrors for adaptive optics, and programmable gratings for spectral tailoring.

 *micromachines*

Editorial

Editorial for the Special Issue on Optical MEMS

Huikai Xie [1,*] and Frederic Zamkotsian [2,*]

1 Department of Electrical and Computer Engineering, University of Florida, Gainesville, FL 32611, USA
2 Aix Marseille Univ, CNRS, CNES, LAM, Laboratoire d'Astrophysique de Marseille, 38 rue Frederic Joliot Curie, 13388 Marseille CEDEX 13, France
* Correspondence: hkxie@ece.ufl.edu (H.X.); frederic.zamkotsian@lam.fr (F.Z.)

Received: 2 July 2019; Accepted: 2 July 2019; Published: 7 July 2019

Optical micro-electro-mechanical systems (MEMS), micro-opto-electro-mechanical systems (MOEMS), or optical microsystems are devices or systems that interact with light through actuation or sensing at a micron or millimeter scale. Optical MEMS have had enormous commercial success in projectors, displays, and fiber optic communications. The best known example is Texas Instruments' digital micromirror devices (DMDs). The development of optical MEMS was impeded seriously by the Telecom Bubble in 2000. Fortunately, DMDs grew their market size even in that economy downturn. Meanwhile, in the last one and half decades, the optical MEMS market has been slowly but steadily recovering. During this time span, the major technological change was the shift of thin-film polysilicon microstructures to single-crystal-silicon microstructures. Especially in the last few years, cloud data centers demand large-port optical cross connects (OXCs), autonomous driving looks for miniature light detection and ranging systems (LiDAR), and virtual reality/augmented reality (VR/AR) demands tiny optical scanners. This is a new wave of opportunities for optical MEMS. Furthermore, several research institutes around the world have been developing MOEMS devices for extreme applications (very fine tailoring of light beam in terms of phase, intensity, or wavelength) and/or extreme environments (high vacuum or cryogenic temperature) for many years.

This special issue contains twelve research papers covering MEMS mirrors [1–10], MEMS variable optical attenuators (VOAs) [11], and tunable spectral filters [12]. These MEMS devices are based on three of the commonly used actuation mechanisms: electrothermal [1], electrostatic [2–7,11], and electromagnetic actuation [8–10]. MEMS optical scanners involving single mirrors are demonstrated or used in [1–3,8–10], while all other optical microsystems employ MEMS mirror arrays that are all based on DMDs [4–7]. This special issue also includes one review paper on metalens-based miniaturized optical systems [13].

Among the papers on single MEMS mirrors, two are focused on MEMS device fabrication [1,10], one on optimization of the driving signals [3], one on applying MEMS mirrors for confocal microscopy [2], and two on using MEMS mirrors to generate structural light patterns for 3D measurement [8,9]. Interestingly, there are several papers reporting various applications of DMDs, including spectral filtering [4], Hadamard spectroscopy [5], wavefront/aberration correction [6], and a tunable fiber laser [7].

In particular, Zhou et al. presented the design, fabrication, and characterization of an electrothermal MEMS mirror with large tip-tilt scan around ±8° and large piston scan of 114 μm at only 2.35 V as well as large resonance frequencies of 1.5 kHz (piston) and 2.7 kHz (tip-tilt); this device survived 220 billion scanning cycles [1]. Lei et al. developed a low-cost FR4-based electromagnetic scanning micromirror integrated with an electromagnetic angle sensor; this MEMS mirror achieved an optical scan angle of 11.2° with a low driving voltage of only 425 mV at resonance (361.8 Hz) [10]. Kim et al. demonstrated an original driving scheme of an electrostatic microscanner in a quasi-static mode based on an input shaping method by an experimental transfer function; the usable scan range was extended up to 90% or higher for most frequencies up to 160 Hz [3].

On the applications side, Yao et al. modified a confocal microscope for including a resonant MEMS scanner in order to miniaturize the system [2]. Hu et al. proposed a new multiple laser stripe scanning profilometry based on a scanning mirror that can project high quality movable laser stripes, delivering high-quality images, mechanical movement noise elimination, and speckle noise reduction [5]. Yang at al. combined the high accuracy of the fringe projection profilometry with the robustness of the laser stripe scanning and demonstrated 3D shape measurement of surfaces with large reflection variations using a biaxial scanning micromirror projection system [9].

Gao et al. showed that a programmable filter based on a DMD can experimentally reach a minimum bandwidth as low as 12.5 GHz in C-band, where the number of channels and the center wavelength can be adjusted independently, as well as the channel bandwidth and the output power [4]. Lu et al. employed a new Hadamard mask of variable-width stripes to improve the Signal-to-Noise Ratio (SNR) of a Hadamard transform near-infrared spectrometer by reducing the influence of stray light [5]. Carmichael Martins et al. confirmed that using a DMD for aperture scanning can perform efficiently to measure ocular aberrations sequentially, even for highly aberrated wavefronts [6]. Li et al. demonstrated a tunable fiber laser with high tuning resolution in the C-band, based on a DMD chip as a programmable wavelength filter, and an echelle grating to achieve high-precision tuning [7].

Finally, Sun et al. were able to reduce the wavelength-dependent loss (WDL) and the polarization-dependent loss (PDL) of MEMS-based variable optical attenuators (VOAs) by using a specific shape of the end-face of the collimating lens [11]. Liu et al. have chosen to use a new dual-mode liquid-crystal (LC) device incorporating a Fabry–Perot cavity and an arrayed LC micro-lens for performing simultaneous electrically adjusted filtering and zooming in the infrared wavelength range by adjusting the transmission spectrum and the point spread function of the incident micro-beams [12]. Li et al. reviewed the use of a metasurface-based flat lens (metalens) for miniaturized optical imaging and sensing systems, especially in the bio-optics field, including a large field of view (FOV), chromatic aberration, and high-resolution imaging [13].

We would like to take this opportunity to thank all the authors for submitting their papers to this Special Issue. We also want to thank all the reviewers for dedicating their time and helping to improve the quality of the submitted papers.

Conflicts of Interest: The authors declare no conflict of interest.

References

1. Zhou, L.; Zhang, X.; Xie, H. An Electrothermal Cu/W Bimorph Tip-Tilt-Piston MEMS Mirror with High Reliability. *Micromachines* **2019**, *10*, 323. [CrossRef] [PubMed]
2. Yao, C.Y.; Li, B.; Qiu, Z. 2D Au-Coated Resonant MEMS Scanner for NIR Fluorescence Intraoperative Confocal Microscope. *Micromachines* **2019**, *10*, 295. [CrossRef] [PubMed]
3. Kim, K.; Moon, S.; Kim, J.; Park, Y.; Lee, J.H. Input Shaping Based on an Experimental Transfer Function for an Electrostatic Microscanner in a Quasistatic Mode. *Micromachines* **2019**, *10*, 217. [CrossRef] [PubMed]
4. Gao, Y.; Chen, X.; Chen, G.; Tan, Z.; Chen, Q.; Dai, D.; Zhang, Q.; Yu, C. Programmable Spectral Filter in C-Band Based on Digital Micromirror Device. *Micromachines* **2019**, *10*, 163. [CrossRef] [PubMed]
5. Lu, Z.; Zhang, J.; Liu, H.; Xu, J.; Li, J. The Improvement on the Performance of DMD Hadamard Transform Near-Infrared Spectrometer by Double Filter Strategy and a New Hadamard Mask. *Micromachines* **2019**, *10*, 149. [CrossRef] [PubMed]
6. Carmichael Martins, A.; Vohnsen, B. Measuring Ocular Aberrations Sequentially Using a Digital Micromirror Device. *Micromachines* **2019**, *10*, 117. [CrossRef] [PubMed]
7. Li, J.; Chen, X.; Dai, D.; Gao, Y.; Lv, M.; Chen, G. Tunable Fiber Laser with High Tuning Resolution in C-band Based on Echelle Grating and DMD Chip. *Micromachines* **2019**, *10*, 37. [CrossRef] [PubMed]
8. Hu, G.; Zhou, X.; Zhang, G.; Zhang, C.; Li, D.; Wang, G. Multiple Laser Stripe Scanning Profilometry Based on Microelectromechanical Systems Scanning Mirror Projection. *Micromachines* **2019**, *10*, 57. [CrossRef] [PubMed]
9. Yang, T.; Zhang, G.; Li, H.; Zhou, X. Hybrid 3D Shape Measurement Using the MEMS Scanning Micromirror. *Micromachines* **2019**, *10*, 47. [CrossRef] [PubMed]

10. Lei, H.; Wen, Q.; Yu, F.; Zhou, Y.; Wen, Z. FR4-Based Electromagnetic Scanning Micromirror Integrated with Angle Sensor. *Micromachines* **2018**, *9*, 214. [CrossRef] [PubMed]
11. Sun, H.; Zhou, W.; Zhang, Z.; Wan, Z. A MEMS Variable Optical Attenuator with Ultra-Low Wavelength-Dependent Loss and Polarization-Dependent Loss. *Micromachines* **2018**, *9*, 632. [CrossRef] [PubMed]
12. Liu, Z.; Chen, M.; Xin, Z.; Dai, W.; Han, X.; Zhang, X.; Wang, H.; Xie, C. Research on a Dual-Mode Infrared Liquid-Crystal Device for Simultaneous Electrically Adjusted Filtering and Zooming. *Micromachines* **2019**, *10*, 137. [CrossRef] [PubMed]
13. Li, B.; Piyawattanametha, W.; Qiu, Z. Metalens-Based Miniaturized Optical Systems. *Micromachines* **2019**, *10*, 310. [CrossRef] [PubMed]

Article

An Electrothermal Cu/W Bimorph Tip-Tilt-Piston MEMS Mirror with High Reliability

Liang Zhou, Xiaoyang Zhang and Huikai Xie *

Department of Electrical and Computer Engineering, University of Florida, Gainesville, FL 32611, USA; l.zhou@ufl.edu (L.Z.); xzhang292@gmail.com (X.Z.)
* Correspondence: hkxie@ece.ufl.edu; Tel.: +1-352-846-0441

Received: 23 April 2019; Accepted: 9 May 2019; Published: 14 May 2019

Abstract: This paper presents the design, fabrication, and characterization of an electrothermal MEMS mirror with large tip, tilt and piston scan. This MEMS mirror is based on electrothermal bimorph actuation with Cu and W thin-film layers forming the bimorphs. The MEMS mirror is fabricated via a combination of surface and bulk micromachining. The piston displacement and tip-tilt optical angle of the mirror plate of the fabricated MEMS mirror are around 114 μm and ±8°, respectively at only 2.35 V. The measured response time is 7.3 ms. The piston and tip-tilt resonant frequencies are measured to be 1.5 kHz and 2.7 kHz, respectively. The MEMS mirror survived 220 billion scanning cycles with little change of its scanning characteristics, indicating that the MEMS mirror is stable and reliable.

Keywords: MEMS mirror; electrothermal bimorph; Cu/W bimorph; electrothermal actuation; reliability

---

1. Introduction

Microelectromechanical (MEMS) mirrors can actively steer light beams. They play an important role in various optical systems and have been widely used in displays [1–3], optical switching [4–6], Fourier transform spectroscopy [7,8], optical endomicroscopy [9–14], tunable lasers [15,16], structured illumination [17], and light detection and ranging (LiDAR) [18,19]. The development of MEMS mirrors dates back to 1980 when Dr. Kurt Petersen published a seminal paper on a torsional mirror using silicon as the mechanical material [20]. Later, in 1987, Dr. Larry Hornbeck at Texas Instruments successfully invented and developed digital micromirror devices that now dominate the projector market [21]. The market size of MEMS mirrors has been growing for decades, and various MEMS mirrors with advanced features for specific applications are still being developed.

Electrostatic, piezoelectric, electromagnetic, and electrothermal actuations have been commonly used in MEMS mirrors [2]. Every actuation mechanism has its advantages and disadvantages. For instance, electrostatic mirrors usually have the advantages of fast response and low power consumption but at the cost of high driving voltage [2]. Due to the large area of comb drives, the fill factor of the active mirror surface is typically low unless a dedicated mirror transfer process is employed [22]. On the other hand, electrothermal MEMS mirrors have large scan angle, low driving voltage, and high fill factor [11,23–26], making them especially suitable for biomedical endoscopic imaging applications.

A variety of MEMS mirrors based on electrothermal bimorph actuators have been reported [23–26]. An electrothermal bimorph comprises two materials with different coefficients of thermal expansion (CTEs), as shown in Figure 1a. If one end of the bimorph is clamped, the other end will curl up or down as the temperature changes. Cr/SiO_2 [27], NiCr/SU-8 [28], Au/Si [29], Al/W [30], and Al/SiO_2 [11,23–26] material pairs heave been used to form bimorph actuators. The Al/SiO_2 pair is used most often because of their large CTE difference and their wide processing availability in almost any MEMS or integrated circuit (IC) fabrication facilities. However, Al is a metal with low melting point (660 °C), and is susceptible to creep failure [31]. SiO_2 is a brittle material, which may result in fracture of bimorphs due to fabrication defects and overstress. Thus, the lifetime and reliability of Al/SiO_2 bimorph based MEMS mirrors may be limited [31].

Therefore, a new material pair for bimorphs is needed to obtain more reliable MEMS mirrors. Cu and W have high Young's moduli, their CTE difference is relatively large, and their thermal diffusivities are also large. Thus, high stiffness and fast thermal response can be expected from Cu/W bimorphs. Zhang et al. demonstrated a Cu/W bimorph based electrothermal MEMS mirror using a lateral-shift-free (LSF) bimorph design [32]. The LSF bimorph actuator consists of three Cu/W bimorph segments (b1, b2, and b3) and two Cu/W/Cu multimorph segments (m1, m2), as shown in the Figure 1b,c. By properly choosing the length ratios of these five segments, the LSF bimorph design minimizes the lateral shift of the central mirror plate. This LSF design also achieves large vertical displacement by utilizing temperature-insensitive Cu/W/Cu multimorphs to amplify the displacement generated from the curling Cu/W bimorphs. An SEM of the LSF Cu/W MEMS mirror is shown in Figure 1d; a large piston displacement of 320 μm and a large scan angle of ±18° were obtained [32]. However, due to the long actuator beams, the stiffness of the bimorph actuators is low (only about 0.1 N/m for the design in Figure 1d) and the thermal resistance is large, causing long thermal response time (the thermal time constant was about 6 ms for the design in Figure 1d). In addition, this LSF MEMS mirror's effective fill factor, that is, the ratio of the area of the mirror plate to the area occupied by both the actuators and mirror plate, is only about 35%. Furthermore, the mirror plate has a small in-plane rotation upon piston actuation because the four actuators are not completely symmetric.

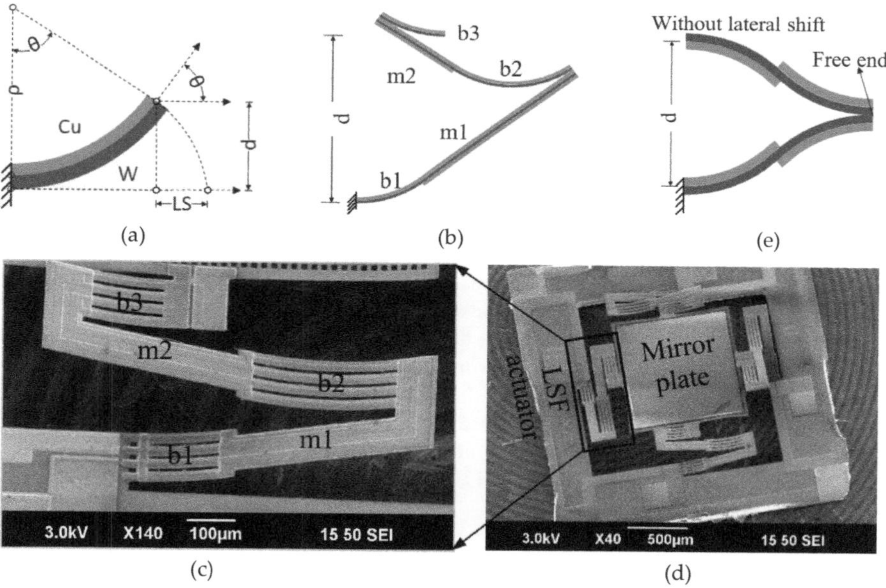

Figure 1. Various bimorph structures: (**a**) A single cantilever bimorph. (**b**) A lateral-shift-free (LSF) bimorph actuator; (**c**) A scanning electron micrograph (SEM) of a Cu/W LSF bimorph actuator; (**d**) An SEM of a Cu/W mirror with LSF design; (**e**) An inverted-series-connected (ISC) bimorph actuator. b1, b2, b3: bimorph segment #1, #2, and #3; m1, m2: multimorph segment #1 and #2.

In this paper, we present a new electrothermal Cu/W bimorph MEMS mirror with an inverted-series-connected (ISC) structure. As shown in Figure 1e, an ISC structure achieves vertical displacement through connecting four segments of bimorphs with flipped layers in series. The ISC actuator design was firstly developed by Todd et al. to overcome the lateral shift and the tip-tilt angle of a single bimorph [25]. Compared to the LSF actuator in Figure 1b, this ISC actuator eliminates the long and wide multimorphs that deteriorates the fill factor and resonant frequency. Thus, this ISC actuator design can increase both stiffness and fill factor. At the same time, this ISC bimorph actuator design can be made completely symmetric, in which every bimorph is the same except the layer sequence. The W layer of the bimorphs also functions as heaters. This concept was initially reported in [33], where a downward ISC Cu/W mirror was reported with preliminary results. This paper focuses on the design, optimization, fabrication, and characterization of an upward ISC Cu/W mirror.

In the following, the bimorph material selection process is discussed in Section 2, device design including structure parameters and simulation is presented in Section 3, the detailed device fabrication process is introduced in Section 4, and the device characterization including quasi-static, dynamic, and long-term stability tests is presented in Section 5.

2. Material Selection

Material selection is crucial to designing reliable bimorphs for MEMS mirrors. Thin film dielectric materials are fragile, so metals are preferred. For metal microstructures, creep and fatigue are among the most important concerns [31]. Alloys are often used over pure metals. For example, Al alloys are successfully used by Texas Instruments to reduce the creep of digital micromirror devices [31]. However, alloys with proper compositions are often difficult to find especially for MEMS processes-compatible ones, so only the pure metals commonly used in MEMS or semiconductor industries, as listed in Table 1, were considered. Evaluation of the stiffness, bimorph responsivity, response time, and maximum working temperature of a single bimorph with two materials of the same width was used to select the two materials.

Table 1. Material properties of commonly used MEMS materials [34].

Material	CTE (10^{-6}/K)	Thermal Conductivity (W/mK)	Young's Modulus (GPa)	Melting Point (°C)	Yield Strength (MPa)
Si	3.0	150.0	179	1414	-
SiO_2	0.4	1.4	70	1700	-
Al	23.6	237.0	70	660	124
Au	14.5	318.0	78	1064	-
Cu	16.9	401.0	120	1083	262
W	4.5	173	410	3410	550
Cr	5.0	93.9	140	1907	200

The tilt angle at the end of the bimorph was determined by the intrinsic stresses and the extrinsic stresses in the two thin-film layers. The intrinsic stresses, which are incurred by the materials and deposition temperature, determined the initial tip-tilt angle and displacement. Miniaturized intrinsic stresses were expected to make the mirror surface at the same level as the substrate, facilitating the fabrication and applications. W was just the right material whose residual stress could be well controlled through adjusting the argon pressure or substrate temperature during sputtering.

The bimorph responsivity, defined as the ratio of the rotation angle at the end of the bimorph, $\Delta\theta$, over the temperature change, ΔT, is expressed as [35]:

$$\Delta\theta/\Delta T = \frac{\beta_b l}{t_a + t_b}(\alpha_a - \alpha_b), \tag{1}$$

where α_a and α_b are the CTE's of material a and b, respectively, t_a and t_b are the thicknesses of material a and b, respectively, β_b is the curvature coefficient of the bimorph, and l is the length of the bimorph. According to Equation (1), the bimorph responsivity is proportional to the CTE difference. Thus, Al and SiO_2 are often chosen as the bimorph materials because of their large CTE difference of 23.2×10^{-6}/K. The CTE difference between Al and W [30] is comparable to that of Al and SiO_2, but Al would incur creep. Although the CTE difference between Cu and W is only around 60% of that of Al and SiO_2, their melting points are much larger than that of Al. Therefore, Cu and W bimorphs can work at higher temperature to achieve similar rotation angles as Al and SiO_2 bimorphs. In addition, creep is smaller for a metal with higher melting temperature, so the creep failure of Cu/W MEMS mirrors will be greatly reduced. Note that the temperature change for a given electrical power is determined by the thermal resistance between the bimorph and the substrate as well as the heat loss to the air via convection; more details can be found in [36].

The equivalent stiffness of the bimorph in Figure 1a can be found as:

$$k = \frac{3EI}{l^3}, \text{ where } EI = \frac{wt_b^3 t_a E_b E_a}{12(t_a E_a + t_b E_b)} K_1, \text{ and } K_1 = 4 + 6\frac{t_a}{t_b} + 4\left(\frac{t_a}{t_b}\right)^2 + \frac{E_a}{E_b}\left(\frac{t_a}{t_b}\right)^3 + \frac{E_b}{E_a}\frac{t_b}{t_a}. \quad (2)$$

Therefore, the stiffness is highly dependent on the Young's moduli of the materials a and b as well as their thickness ratio. The equivalent rigidity of a Cu/W bimorph with a same width and equivalent thickness is around three times of that of an Al/SiO_2 bimorph. In other words, compared to an Al/SiO_2 bimorph, a Cu/W bimorph with a much smaller thickness can be used to achieve the same stiffness.

From a simplified one-dimensional thermal lumped model, the thermal response time is inversely proportional to the thermal diffusivity (α), that is,

$$t \propto R_{th} C_{th} \propto \frac{1}{k/\rho c_p} = \frac{1}{\alpha}, \quad (3)$$

in which thermal convection and radiation are neglected for simplification. The thermal diffusivities of Cu and W are comparable to Al (120% and 70% of that of Al, respectively), but over 75 times higher than that of SiO_2. Therefore, the thermal response time of a Cu/W bimorph will be much smaller than that of an Al/SiO_2 bimorph.

For a reliable bimorph, the materials must work in the region of elasticity, that is, the maximum bending stress must not exceed the yield strength. According to Table 1, the yield strengths of Cu and W are about two times and four times higher than that of Al, respectively. In addition, Cu and W are widely available for micromachining and their fabrication processes are mature. Also, W, whose resistivity is 5.6×10^{-8} Ω·m, is commonly used in incandescent light bulbs. Therefore, the W layer can function as a heater.

With all the above merits, Cu and W were selected as the materials for making the ISC bimorph actuators in this work.

3. Device Design

The schematic of the MEMS mirror built on the Cu/W bimorphs with ISC structures is shown in Figure 2. The central mirror plate was made of a 20-μm-thick silicon for optical flatness and a 0.2-μm-thick aluminum on the surface for high reflectance. The mirror plate was 1 mm in diameter and suspended by four pairs of ISC actuators. There were thin silicon oxide beams between the ISC actuators and the mirror plate, as shown in the inset of Figure 2, functioning as thermal isolation to confine the Joule heat to the bimorphs and minimize the temperature rise on the mirror plate. There was another set of silicon oxide beams between the bimorph actuators and the substrate, forming a thermal barrier to reduce the heat to the silicon substrate. There were eight pads extended from the tungsten layer of the bimorphs with two pads on each side of the substrate. Thus, every actuator could be actuated separately. The mirror plate could move vertically when all the actuators were applied with the same voltage, and could rotate when the four actuators were applied with different voltages.

If the W and Cu layers have the same width, the optimal thickness ratio of these two layers is 0.56 for achieving the maximum displacement [23]. However, the actual widths of the Cu and W layers were different. For the sake of good step coverage and reliable photolithography, the Cu layers were chosen to be wider than the W layers. In this design, the width of the Cu layer was set as 30 μm, while the W width was set as 16 μm, which ensured the W layer was either fully covered by the Cu layer or fully on top of the Cu layer, even with minor mask aligning errors. The structure parameters of the Cu/W ISC MEMS mirror are given in Table 2. With the aid of COMSOL simulation, it was found that the optimal W-to-Cu thickness ratio is 0.77. By considering the required robustness of the MEMS mirror and easy fabrication, the actual W and Cu thicknesses were chosen as 1.0 μm and 1.3 μm, respectively. The flexural rigidity (EI) of the Cu/W bimorph was 4.82 Pa·mm^4, which is 4.25 times that of the Al/SiO_2 bimorph in [3] whose Al and SiO_2 thicknesses were 1.1 μm and 1.2 μm, respectively.

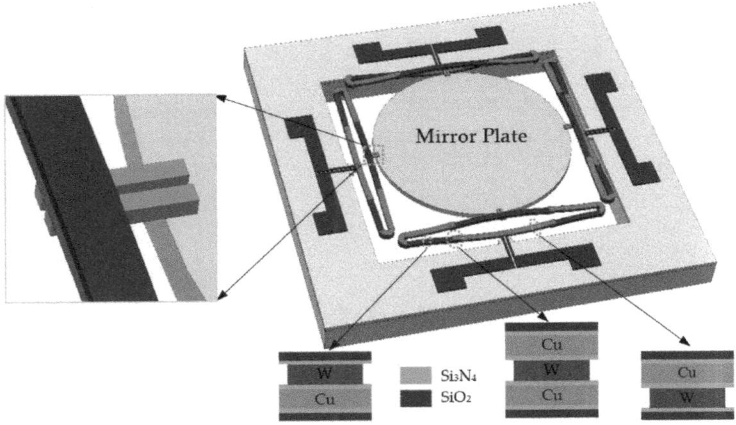

Figure 2. Schematic of a microelectromechanical (MEMS) mirror based on Cu/W ISC actuators.

Table 2. Design parameters of Cu/W MEMS Mirror.

Structure Parameters	Value
Device footprint	2.2 mm × 2.2 mm
Diameter of the mirror plate	1 mm
Mirror plate thickness	20 μm
Length of each bimorph	180 μm
Width of W	16 μm
Width of Cu	30 μm
Length of overlap	60 μm

Note that even when Cu is passivated with SiO_2, oxygen can still diffuse through and oxidize Cu, so a thin layer (~50 nm) of Si_3N_4 is needed as a diffusion barrier layer to wrap Cu layers. A finite element 3D model with the parameters as shown in Table 2 was created in COMSOL Multiphysics (version 5.4, COMSOL Inc., Stockholm, Sweden) to show the performance of the MEMS mirror. All layers including the dielectric layers were considered. As shown in Figure 3, the first and second resonant frequencies were 1.493 kHz and 2.518 kHz, respectively. Since the mass of the mirror plate was several orders of magnitude larger than those of the actuators, the stiffness of a single actuator can be calculated by:

$$k_{\text{act}} = \frac{1}{4} m_{\text{plate}} (2\pi f_p)^2, \tag{4}$$

where m_{plate} is the mass of the mirror plate, and f_p is the piston resonant frequency. Thus, the stiffness of one double S-shaped bimorph actuator is $k_{\text{act}} = 0.81\ N/m$. Tip-tilt actuation can be realized by applying different temperature at the four different actuators. Raising the same temperature on four actuators at the same time results in a piston movement.

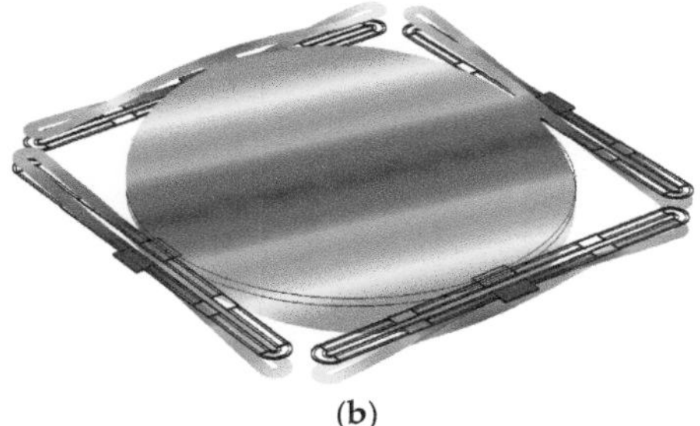

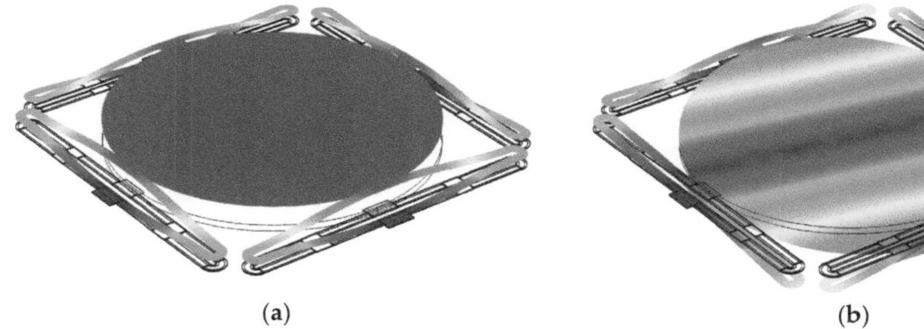

Figure 3. The modal simulation of the Cu/W mirror. (**a**) First resonant mode, piston, at the frequency of 1.493 kHz; (**b**) Second resonant mode, tip-tilt, at the frequency of 2.518 kHz.

4. Device Fabrication

The mirror pate and the bimorphs were released by bulk micromachining. The fabrication process flow is illustrated in Figure 4. First, a 1-μm-thick plasma enhanced chemical vapor deposition (PECVD) SiO_2 was deposited on a 4″ silicon on insulator (SOI) wafer and wet etched to form electrical insulation on top of the silicon device layer and thermal isolation from the bimorphs to the substrate and to the mirror plate (Figure 4a). A 0.15/0.05-μm PECVD SiO_2/Si_3N_4 was deposited and reactive-ion-etch (RIE) patterned as the bottom diffusion barrier layer of the bimorphs. A 1.3 μm Cu layer was sputtered and lift-off to define the bimorphs that require Cu as the bottom layer (Figure 4b). A 0.1 μm Si_3N_4 layer was deposited and RIE patterned for electrical isolation and a 1 μm W layer was sputtered and patterned via lift-off to define the bimorphs; the W layer also worked as the resistor for Joule heating (Figure 4c). Another 0.1 μm Si_3N_4 layer was deposited on top of the W layer and vias were opened on top of W by RIE. The second Cu layer was then sputtered and lift-off to define the bimorphs that required Cu as the top layer, followed by another 0.05/0.15 μm thin PECVD Si_3N_4/SiO_2 deposited as the diffusion barrier layer of the bimorphs (Figure 4d). These Si_3N_4/SiO_2 multilayer dielectric layers between the bimorphs were etched by RIE for later release, and vias were formed on top of the Cu layer. A 0.5 μm Al layer was sputtered to define the mirror surface and the bonding pads on top of Cu (Figure 4e). At this point, all the processes on the front side of the wafer were done.

After the wafer was flipped over, a 0.2 μm SiO_2 was deposited and RIE etched to define the regions corresponding to the bimorphs (Figure 4f). Next, a photoresist pattern corresponding to the entire bimorphs plus the mirror plate was formed (Figure 4g). Then, a first round of deep reactive-ion-etching (DRIE) was used to etch trenches into the silicon substrate by about 40 μm while the silicon under the mirror plate was still intact (Figure 4h). Next, the 0.2 μm SiO_2 was removed by RIE. A second round of DRIE was used to etch down to the buried oxide (BOX) layer (Figure 4i), followed by removing the BOX layer with RIE (Figure 4j). Finally, a third round of DRIE was done to remove all the remaining silicon layer under the bimorphs and the mirror plate (Figure 4k). As the front-side Al mirror surface was not exposed to DRIE, the surface quality of the mirror was high.

An SEM of a fabricated ISC MEMS mirror is shown in Figure 5. The device footprint was 2.2 mm × 2.2 mm with an effective fill factor of 48% even with a circular mirror plate, which was still 37% larger than the LSF design in Figure 1d. The initial elevation of the mirror plate was measured to be 128 μm, incurred by the residual stresses, and the bimorphs were free of oxidation. The measured resistances of the four actuators were 30.4–32.4 Ω at room temperature.

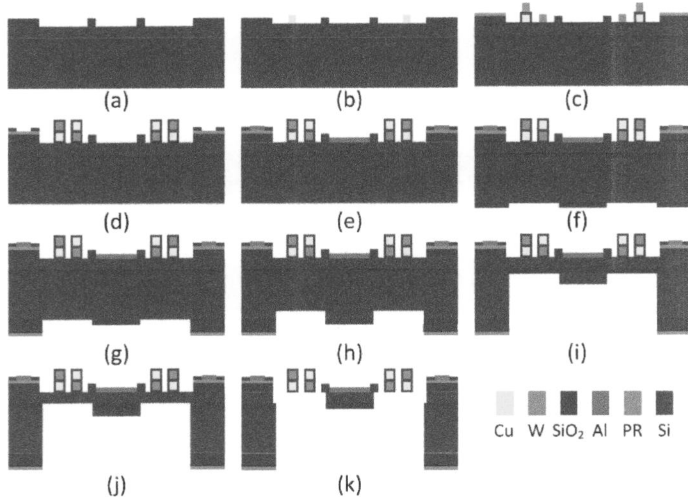

Figure 4. Fabrication process flow of the MEMS mirror.

Figure 5. SEM of a fabricated MEMS mirror.

5. Characterization

The quasi-static, dynamic and frequency responses of the MEMS mirror were characterized. The long-term reliability was also tested. These experimental results are presented below.

5.1. Static Response

When a same direct current (DC) voltage was applied to all four actuators, the mirror plate moved vertically. An optical microscope was used to measure the heights of the mirror plate at different DC voltages. Figure 6 plots the piston displacements of the mirror plate versus the applied DC voltage and the corresponding power, respectively, showing that the mirror plate traveled 114 μm at only 2.35 V or 475 mW. When a voltage was applied on one actuator while leaving other three actuators open-circuit, the mirror plate tilted. Figure 7 shows the optical scan angle of the mirror plate versus the applied DC voltage, showing that the mirror plate tilted 4° (or 8° optical angle) at 2.35 V. Also plotted on Figure 7 is the displacement of the mirror plate center reaching to 51 μm at 2.35 V. This means that the mirror plate was flipping instead of rotating along its central axis. In order to keep the center stationary, a differential drive—that is, applying a pair of differential voltages on one pair of opposing actuators with a DC bias set on all four actuators—can be used [23].

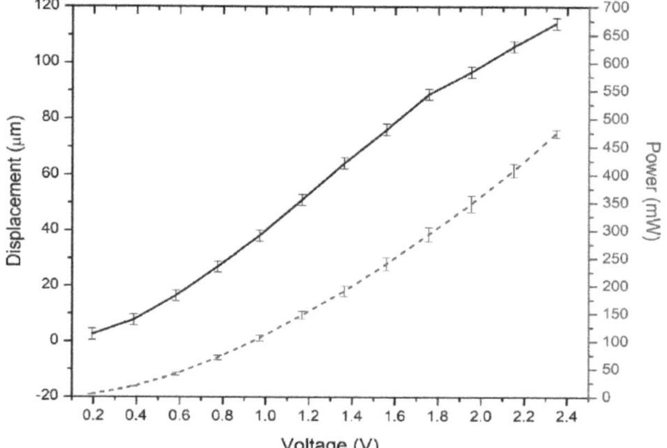

Figure 6. The vertical displacement (solid line), and the corresponding consumed power (dash line) versus the applied voltage. The errors for the displacement measurement were about ± 2 μm resulting from the errors of the microstage position reading and the focal point determination of the optical microscope.

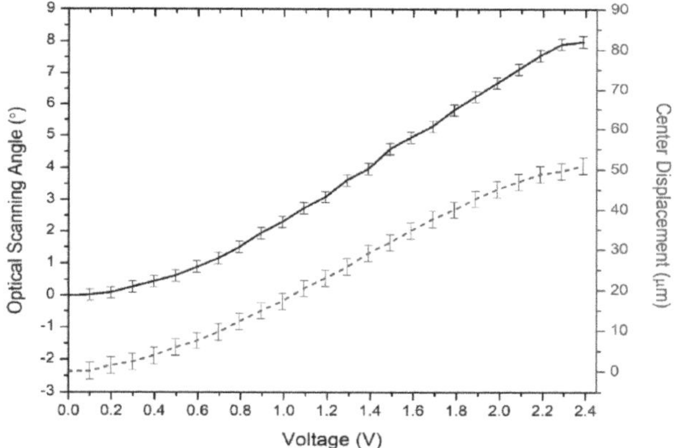

Figure 7. The optical scan angle (solid line), and the corresponding center displacement (dash line) versus the applied voltage.

5.2. Frequency Response

The frequency response was measured by using a network analyzer. An input sweep-frequency voltage signal, $1 + 0.1 \times \cos(2\pi ft)$ V, generated by the network analyzer, was applied to the MEMS mirror; the laser spot reflected by the scanning mirror plate was picked up by a photosensitive detector (PSD) whose output signal was a measure of the tilt angle of the mirror plate. This signal was sent back to the network analyzer, so the frequency response was directly obtained. The measured frequency response is shown in Figure 8, where the first mode (piston) was 1.55 kHz and the second mode (tip-tilt) was 2.7 kHz. The resonant modes were well predicted by the simulation results (see Figure 3) with an error less than 7%. The Q factor of the tip-tilt mode was 25.5. The 3 dB cutoff frequency, f_{3dB}, was around 50 Hz, which was the result of the thermal response. Thus the thermal time constant $\tau_T = \frac{1}{2\pi f_{3dB}} = 3.2$ ms, which is around 50% of the time of the LSF actuator as shown in Figure 1d [32].

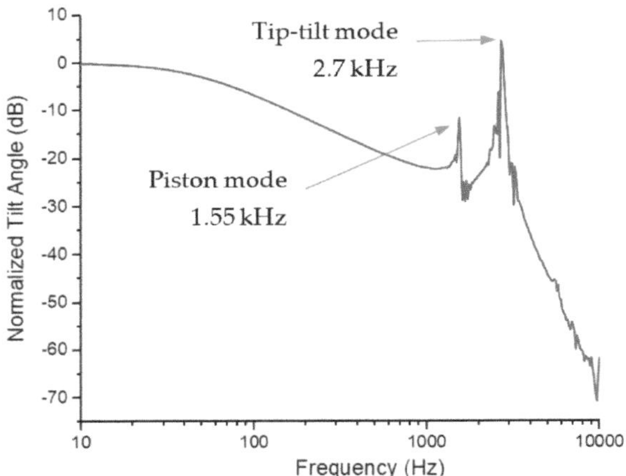

Figure 8. The frequency response of the micromirror from 1 Hz to 10 kHz.

Note that the measured piston resonant frequency was f_p = 1550 Hz, so $k_{act} = \frac{1}{4} m_{plate} (2\pi f_p)^2 =$ 0.87 N/m, which is in a good agreement with the simulated stiffness and seven times as much as that of the design in Figure 1d. According to Figure 8, the measured displacement was 114 μm at 2.35 V; thus, the force generated by each electrothermal bimorph actuator was about 92.3 μN at that voltage. Therefore, this type of electrothermal actuators is also suitable for applications that require relatively large driving force, such as MEMS lens scanners.

5.3. Step Response

Step response and frequency response have been characterized. The step response was measured with a laser shinning on the mirror plate and a PSD detecting the position of the reflected laser beam. A 10 Hz square wave with a 50% duty cycle and a 1V amplitude was used to drive a single actuator. The result is shown in Figure 9a, with the zoom-in views of the rise and fall response in Figure 9b,c, respectively. The 10–90% rise and fall time were 7.3 ms and 8.5 ms, respectively. The fall time was 1.2 ms or 16% longer that the rise time, which is believed to be caused by the heat stored inside the mirror plate flowing back into the actuators during cooling. Note that in Figure 9 the step response was smooth with no overshoot and very small ringing. This was the result of the low-pass filtering effect of the thermal response. There is a 20 dB/Decade roll-off after the thermal cut-off frequency. Thus, the resonant peak height was proportional to the thermal response time and inversely proportional to the resonant frequency. According to the finding reported in [37], the overshoot and ringing of a step response of an electrothermal actuator will be greatly suppressed if $\tau_T \cdot f_0 > Q/(2\pi)$. In this case, $\tau_T \cdot f_{tip-tilt} = 3.2 \times 2.7 = 8.6$, which is greater than $Q/2\pi = 4$.

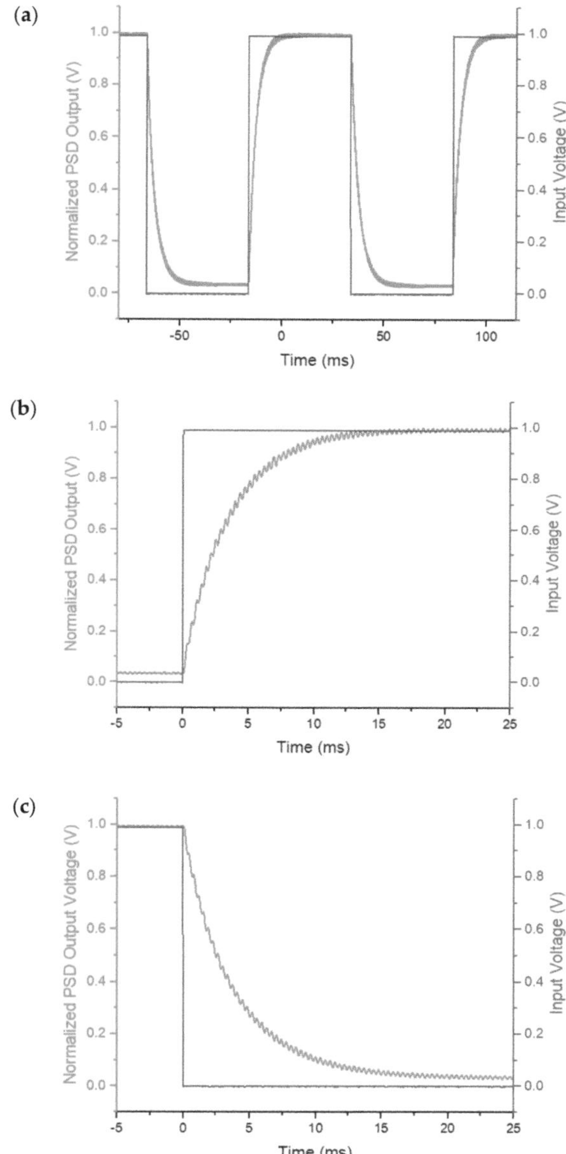

Figure 9. (**a**) Step response of one actuator of the MEMS Mirror; (**b**) zoom-in rise edge; (**c**) zoom-in fall edge.

5.4. Reliability

The reliability was characterized by recording the tip-tilt resonant frequency and its corresponding scan angle for more than 30 months. A sine waveform voltage signal with an amplitude of 1 V and an offset of 1 V was applied to one actuator of the fabricated Cu/W ISC MEMS mirror. A laser beam was shined on the mirror plate, and the reflected laser beam was projected to a screen. The driving frequency was continuously adjusted to its resonant frequency to keep the scan angle at its maxima, and both the frequency and the corresponding scan angle were recorded. The resonant frequency and the resonant scan angle over time are plotted in Figure 10. During the first 9 months, the resonant frequency dropped slowly from 2718 Hz to 2705 Hz, corresponding to a frequency shift as small as 0.47%. The resonant scan angle decreased gradually during the first 13 months, dropping about 10.7% (from 53.2° to 47.5°). After 13 months' continuous running, both the resonant frequency and the scan angle became very stable, maintaining at 2705 ± 1 Hz and 47 ± 0.4°, respectively. Thus, this MEMS mirror is reliable and can be used for long-term operation and even for open-loop driving. Note that even though the overall changes of the resonant frequency and maximum scan angle were small, there still existed a rapid-changing time period, which happened to be the beginning part of the device operation life time. This rapid-changing period is believed to be undergoing the burn-in process. More experiments are needed to verify this.

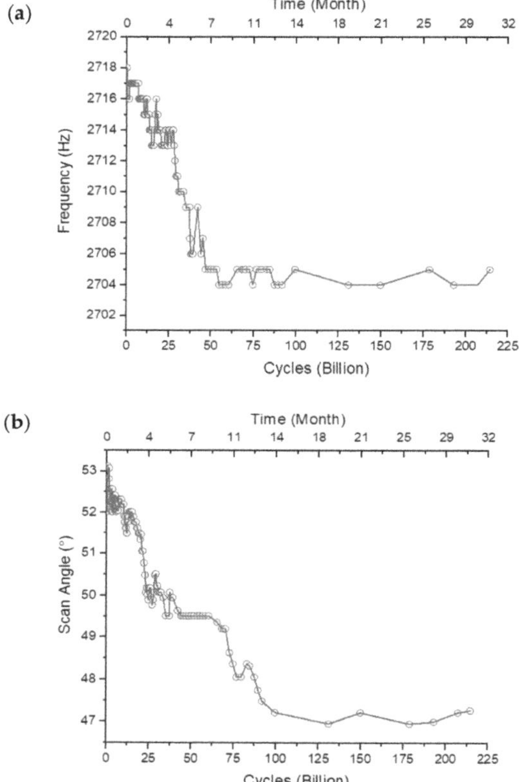

Figure 10. (**a**) Long-term frequency shift; (**b**) long-term scan angle change at the corresponding tip-tilt resonant frequency.

Figure 11a shows a scanning electron microscope (SEM) image of the MEMS mirror after the long-term (30 months) test. The entire device still looked clean except the joint part between the mirror plate and the actuated bimorph actuator was contaminated. As shown in the close-view images of the actuated bimorphs (Figure 11b,c), some extrusion accumulated at the edge of the bimorphs. This is believed to be copper oxide. Since the Si_3N_4 barrier layer was partially etched during DRIE, some copper atoms diffused out and were oxidized on the surface of the bimorphs. It is also believed that this Cu oxidization accounted for the relatively large changes of the resonant frequency and the scan angle during the initial 13-month long-term test. More study will be performed to understand the observed phenomena.

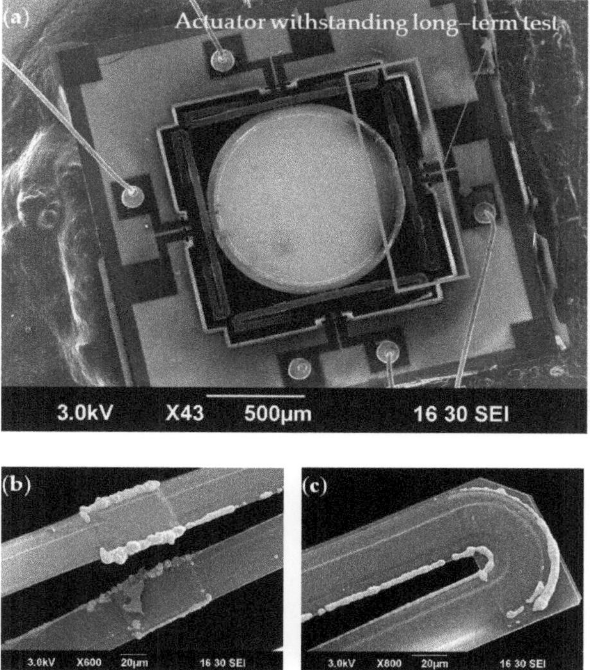

Figure 11. SEM pictures of the MEMS mirror after long-term actuation under the tip-tilt resonant frequency. (**a**) The full device; (**b**) a zoom-in SEM picture of the actuated bimorph near overlap between inversed bimorphs; (**c**) a zoom-in SEM picture of the actuated bimorph near a corner.

6. Conclusions

A reliable MEMS mirror based on the ISC Cu/W bimorph actuator design has been successfully demonstrated. Compared to its LSF bimorph design counterpart with the same dimensions, this ISC design is completely symmetric with six times higher stiffness, 37% higher fill factor, and 50% smaller thermal response time. In addition, the ISC bimorph MEMS mirror has been proven to be stable long-term, surviving over 200 billion cycles of large angular scanning. In the future, this Cu/W ISC design will be further optimized, including eliminating the Cu oxidation by adopting new diffusion barrier layers, reducing the thicknesses of the insulation layers, minimizing the width difference between the Cu and W layers, and increasing the bimorph width to achieve even larger scan angle, higher robustness, and better reliability.

Author Contributions: L.Z. designed and performed the experiments, analyzed the data, and wrote the paper. X.Z. designed and fabricated the MEMS mirror and analyzed the data. H.X. conceived and supervised the design and the experiments and revised the paper critically.

Funding: This work was supported by the National Science Foundation under the award #1512531 and the National Institutes of Health under the award #R01EB020601.

Acknowledgments: Device fabrication was done in the Nanoscale Research Facility of the University of Florida.

Conflicts of Interest: The authors declare no conflict of interest.

References

1. Urey, H. Torsional MEMS scanner design for high-resolution scanning display systems. In Proceedings of the International Symposium on Optical Science and Technology, Seattle, WA, USA, 7–11 July 2002; pp. 27–38.
2. Holmstrom, S.T.; Baran, U.; Urey, H. MEMS laser scanners: A review. *J. Microelectromech. Syst.* **2014**, *23*, 259–275. [CrossRef]
3. Van Kessel, P.F.; Hornbeck, L.J.; Meier, R.E.; Douglass, M.R. A MEMS-based projection display. *Proc. IEEE* **1998**, *86*, 1687–1704. [CrossRef]
4. Aksyuk, V.; Pardo, F.; Carr, D.; Greywall, D.; Chan, H.; Simon, M.; Gasparyan, A.; Shea, H.; Lifton, V.; Bolle, C. Beam-steering micromirrors for large optical cross-connects. *J. Lightwave Technol.* **2003**, *21*, 634–642. [CrossRef]
5. Lee, S.-S.; Huang, L.-S.; Kim, C.-J.; Wu, M.C. Free-space fiber-optic switches based on MEMS vertical torsion mirrors. *J. Lightwave Technol.* **1999**, *17*, 7–13.
6. Stepanovsky, M. A Comparative Review of MEMS-based Optical Cross-Connects for All-Optical Networks from the Past to the Present Day. *IEEE Commun. Surv. Tutor.* **2019**. [CrossRef]
7. Antila, J.; Tuohiniemi, M.; Rissanen, A.; Kantojärvi, U.; Lahti, M.; Viherkanto, K.; Kaarre, M.; Malinen, J. MEMS- and MOEMS-Based Near-Infrared Spectrometers. In *Encyclopedia of Analytical Chemistry*; Wiley: Hoboken, NJ, USA, 2014.
8. Han, F.; Wang, W.; Zhang, X.; Xie, H. Miniature Fourier transform spectrometer with a dual closed-loop controlled electrothermal micromirror. *Opt. Express* **2016**, *24*, 22650–22660. [CrossRef] [PubMed]
9. Sun, J.; Xie, H. MEMS-based endoscopic optical coherence tomography. *Int. J. Opt.* **2011**, *2011*, 825629. [CrossRef]
10. Lu, C.D.; Kraus, M.F.; Potsaid, B.; Liu, J.J.; Choi, W.; Jayaraman, V.; Cable, A.E.; Hornegger, J.; Duker, J.S.; Fujimoto, J.G. Handheld ultrahigh speed swept source optical coherence tomography instrument using a MEMS scanning mirror. *Biomed. Opt. Express* **2014**, *5*, 293–311. [CrossRef]
11. Liu, L.; Wang, E.; Zhang, X.; Liang, W.; Li, X.; Xie, H. MEMS-based 3D confocal scanning microendoscope using MEMS scanners for both lateral and axial scan. *Sens. Actuators Phys.* **2014**, *215*, 89–95. [CrossRef]
12. Piyawattanametha, W.; Patterson, P.; Su, G.; Toshiyoshi, H.; Wu, M. A MEMS non-interferometric differential confocal scanning optical microscope. In *Transducers' 01 Eurosensors XV*; Springer: New York City, NY, USA, 2001; pp. 590–593.
13. Piyawattanametha, W.; Cocker, E.D.; Burns, L.D.; Barretto, R.P.; Jung, J.C.; Ra, H.; Solgaard, O.; Schnitzer, M.J. In vivo brain imaging using a portable 2.9 g two-photon microscope based on a microelectromechanical systems scanning mirror. *Opt. Lett.* **2009**, *34*, 2309–2311. [CrossRef]
14. Zong, W.; Wu, R.; Li, M.; Hu, Y.; Li, Y.; Li, J.; Rong, H.; Wu, H.; Xu, Y.; Lu, Y. Fast high-resolution miniature two-photon microscopy for brain imaging in freely behaving mice. *Nat. Methods* **2017**, *14*, 713–719. [CrossRef]
15. Zhang, X.; Liu, A.; Tang, D.; Lu, C. Discrete wavelength tunable laser using microelectromechanical systems technology. *Appl. Phys. Lett.* **2004**, *84*, 329–331. [CrossRef]
16. Paterson, A.; Bauer, R.; Lubeigt, W.; Uttamchandani, D. Q-switched tunable solid-state laser using a MOEMS mirror. In Proceedings of the 2017 International Conference on Optical MEMS and Nanophotonics (OMN), Santa Fe, NM, USA, 13–17 August 2017; pp. 1–2.
17. Zhang, X.; Koppal, S.J.; Zhang, R.; Zhou, L.; Butler, E.; Xie, H. Wide-angle structured light with a scanning MEMS mirror in liquid. *Opt. Express* **2016**, *24*, 3479–3487. [CrossRef]
18. Kasturi, A.; Milanovic, V.; Atwood, B.H.; Yang, J. UAV-borne lidar with MEMS mirror-based scanning capability. In Proceedings of the Laser Radar Technology and Applications XXI, Baltimore, MD, USA, 19–20 April 2016; Volume 98320.
19. Ito, K.; Niclass, C.; Aoyagi, I.; Matsubara, H.; Soga, M.; Kato, S.; Maeda, M.; Kagami, M. System Design and Performance Characterization of a MEMS-Based Laser Scanning Time-of-Flight Sensor Based on a 256×64-pixel Single-Photon Imager. *IEEE Photonics J.* **2013**, *5*, 6800114. [CrossRef]
20. Petersen, K.E. Silicon torsional scanning mirror. *IBM J. Res. Dev.* **1980**, *24*, 631–637. [CrossRef]
21. Hornbeck, L.J. The DMD^{TM} projection display chip: a MEMS-based technology. *Mrs Bull.* **2001**, *26*, 325–327. [CrossRef]

22. Jung, I.W.; Krishnamoorthy, U.; Solgaard, O. High fill-factor two-axis gimbaled tip-tilt-piston micromirror array actuated by self-aligned vertical electrostatic combdrives. *J. Microelectromech. Syst.* **2006**, *15*, 563–571. [CrossRef]
23. Jia, K.; Pal, S.; Xie, H. An electrothermal tip–tilt–piston micromirror based on folded dual S-shaped bimorphs. *J. Microelectromech. Syst.* **2009**, *18*, 1004–1015.
24. Wu, L.; Xie, H. A large vertical displacement electrothermal bimorph microactuator with very small lateral shift. *Sens. Actuators Phys.* **2008**, *145*, 371–379. [CrossRef]
25. Todd, S.T.; Jain, A.; Qu, H.; Xie, H. A multi-degree-of-freedom micromirror utilizing inverted-series-connected bimorph actuators. *J. Opt. Pure Appl. Opt.* **2006**, *8*, S352. [CrossRef]
26. Fu, L.; Jain, A.; Xie, H.; Cranfield, C.; Gu, M. Nonlinear optical endoscopy based on a double-clad photonic crystal fiber and a MEMS mirror. *Opt. Express* **2006**, *14*, 1027–1032. [CrossRef] [PubMed]
27. Lammel, G.; Schweizer, S.; Renaud, P. *Optical Microscanners and Microspectrometers Using Thermal Bimorph Actuators*; Springer Science & Business Media: Berlin/Heidelberg, Germany, 2013; Volume 14.
28. Kiuchi, Y.; Taguchi, Y.; Nagasaka, Y. Fringe-tunable electrothermal Fresnel mirror for use in compact and high-speed diffusion sensor. *Opt. Express* **2017**, *25*, 758–767. [CrossRef]
29. Morrison, J.; Imboden, M.; Little, T.D.; Bishop, D. Electrothermally actuated tip-tilt-piston micromirror with integrated varifocal capability. *Opt. Express* **2015**, *23*, 9555–9566. [CrossRef]
30. Pal, S.; Xie, H. Fabrication of robust electrothermal MEMS devices using aluminum–tungsten bimorphs and polyimide thermal isolation. *J. Micromech. Microeng.* **2012**, *22*, 115036. [CrossRef]
31. Van Spengen, W.M. MEMS reliability from a failure mechanisms perspective. *Microelectron. Reliab.* **2003**, *43*, 1049–1060. [CrossRef]
32. Zhang, X.; Zhou, L.; Xie, H. A fast, large-stroke electrothermal MEMS mirror based on Cu/W bimorph. *Micromachines* **2015**, *6*, 1876–1889. [CrossRef]
33. Zhang, X.; Li, B.; Li, X.; Xie, H. A robust, fast electrothermal micromirror with symmetric bimorph actuators made of copper/tungsten. In Proceedings of the 2015 Transducers-2015 18th International Conference on Solid-State Sensors, Actuators and Microsystems (TRANSDUCERS), Anchorage, AK, USA, 21–25 June 2015; pp. 912–915.
34. Pal, S.; Zhou, L.; Zhang, X.; Xie, H. Electrothermally actuated MEMS mirrors: Design, modeling, and applications. In *Optical MEMS, Nanophotonics, and Their Applications*; CRC Press: Boca Raton, FL, USA, 2017; pp. 173–200.
35. Chu, W.-H.; Mehregany, M.; Mullen, R.L. Analysis of tip deflection and force of a bimetallic cantilever microactuator. *J. Micromech. Microeng.* **1993**, *3*, 4. [CrossRef]
36. Todd, S.T.; Xie, H. An electrothermomechanical lumped element model of an electrothermal bimorph actuator. *J. Microelectromechan. Syst.* **2008**, *17*, 213–225. [CrossRef]
37. Li, M.; Chen, Q.; Liu, Y.; Ding, Y.; Xie, H. Modelling and Experimental Verification of Step Response Overshoot Removal in Electrothermally-Actuated MEMS Mirrors. *Micromachines* **2017**, *8*, 289. [CrossRef]

Article

2D Au-Coated Resonant MEMS Scanner for NIR Fluorescence Intraoperative Confocal Microscope

Cheng-You Yao [1,2,†], Bo Li [2,3,†] and Zhen Qiu [1,2,3,*]

1 Department of Biomedical Engineering, Michigan State University, East Lansing, MI 48823, USA; yaochen5@msu.edu
2 Institute for Quantitative Health Science and Engineering, Michigan State University, East Lansing, MI 48823, USA; libo2@msu.edu
3 Department of Electrical and Computer Engineering, Michigan State University, East Lansing, MI 48823, USA
* Correspondence: qiuzhen@egr.msu.edu; Tel.: +1-517-884-6942
† These authors contributed equally to this work.

Received: 30 March 2019; Accepted: 26 April 2019; Published: 30 April 2019

Abstract: The electrostatic MEMS scanner plays an important role in the miniaturization of the microscopic imaging system. We have developed a new two-dimensional (2D) parametrically-resonant MEMS scanner with patterned Au coating (>90% reflectivity at an NIR 785-nm wavelength), for a near-infrared (NIR) fluorescence intraoperative confocal microscopic imaging system with a compact form factor. A silicon-on-insulator (SOI)-wafer based dicing-free microfabrication process has been developed for mass-production with high yield. Based on an in-plane comb-drive configuration, the resonant MEMS scanner performs 2D Lissajous pattern scanning with a large mechanical scanning angle (MSA, ±4°) on each axis at low driving voltage (36 V). A large field-of-view (FOV) has been achieved by using a post-objective scanning architecture of the confocal microscope. We have integrated the new MEMS scanner into a custom-made NIR fluorescence intraoperative confocal microscope with an outer diameter of 5.5 mm at its distal-end. Axial scanning has been achieved by using a piezoelectric actuator-based driving mechanism. We have successfully demonstrated ex vivo 2D imaging on human tissue specimens with up to five frames/s. The 2D resonant MEMS scanner can potentially be utilized for many applications, including multiphoton microendoscopy and wide-field endoscopy.

Keywords: resonant MEMS scanner; electrostatic; parametric resonance; NIR fluorescence; intraoperative microscope; 2D Lissajous; fluorescence confocal

---

1. Introduction

The intraoperative microscope has become an emerging bio-imaging technology for clinical applications, including molecular imaging-guided surgery [1]. Other intraoperative imaging tools have been successfully demonstrated, such as wide-field fluorescence [2], confocal [3,4], optical coherence tomography (OCT) [5,6], multiphoton [7,8], etc. Among these state-of-the-art optical imaging modalities, miniaturized fluorescence confocal microscopy holds the promise for many translational applications [3,9], including both early cancer detection and tumor margin delineation. MEMS technology plays an important role in the instrument miniaturization of the intraoperative confocal microscopes, in which MEMS scanners and actuators perform beam steering and focus tuning [10,11]. Various MEMS-enabled confocal microscopes [12–16] have been previously developed by integrating custom-made micro-scanners based on different working principles, such as electromagnetic [17,18], electro-thermal [19–21], electrostatic [22], and thin-film piezoelectric [23,24]. Due to their fast-speed scanning up to large angles in a very small footprint, electrostatic micro-scanners have been widely utilized in MEMS-based microscopes [25–29]. However, to date, only a handful of

MEMS-based near-infrared (NIR) (>785 nm) fluorescence intraoperative confocal microscopes have been demonstrated [9]. Most of the existing intraoperative confocal microscopes perform either the reflective-mode imaging or visible-range fluorescence imaging. Although it has been used for clinical trials [3], the commercial Cellvizio™ (Mauna Kea Tech, Paris, France) intraoperative confocal microscope has a limited field-of-view (FOV) (<400 μm) without *Z*-axis scan. It uses fiber bundles combined with micro-optics, while bulky galvanometer scanners are utilized in a pre-objective way. It only operates in the visible range (488–640 nm) because the fiber bundles have a low transmission efficiency in the NIR range. One of the greatest challenges in the compact NIR fluorescence intraoperative confocal microscope is collecting enough signal to achieve an adequate signal-to-noise ratio (SNR) since the fluorescence emission signal is weak during in vivo imaging. Therefore, the reflectance efficiency of MEMS scanners has to be sufficiently high to ensure efficient laser excitation and fluorescence collection. Aluminum coating (on the full wafer) without pattern has been commonly used [8,15]. In addition, to avoid creating a short-circuit between comb-drive fingers, only a very thin layer of aluminum coating (less than 50 nm) can be used. The thin aluminum coating layer's reflectivity will not be sufficient for weak NIR fluorescence detection from tissue specimens. In the NIR range, the Au coating will provide much better reflectivity (>90%), while the reflectivity efficiency of the aluminum coating is relatively low (80%). Unfortunately, very few Au-coated electrostatic MEMS scanners have been demonstrated or mass-produced for custom-made miniature NIR fluorescence intraoperative confocal microscopes. In this project, based on the parametric resonance working principle [30,31], an electrostatic comb-drive-actuated, gimbal frame-based 2D resonant MEMS scanner has been developed and fully integrated into a newly-developed miniature NIR fluorescence intraoperative confocal microscope (outer diameter (OD) 5.5 mm). To achieve a high reflectivity in the NIR range, the new scanner has been coated with a patterned Au/Ti (Ti: adhesion, low stress) coating layer. Resonant scanners offer large tilting angles with a relatively low driving voltage, compared to the conventional electrostatic MEMS scanners based on staggered or angular vertical comb-drives (SVC or AVC) [8,15,22], which usually require a high driving voltage (>100 V, not safe for humans) for large tilting angles in the DC mode with a raster scanning pattern. By taking advantage of the new resonant scanner in our post-objective scanning-based optics, we are able to realize a large FOV (up to 1000 μm) with very low driving voltage (36 V), which is important for in vivo imaging of humans. Our design was inspired by the seminal work [31] by the team led by Schenk. Unfortunately, the former processes are not suitable for our design. A resonant scanner with a large fill-in factor (on the *X*-axis, "dumb-bell" shape) is required to fit the post-objective scanning-based optics for depth imaging with a large FOV. For example, in the former process, KOH-based wet etching on the backside of the wafer required more supporting materials on both the device and handle silicon layers. In our study, based on a single silicon-on-insulator (SOI) wafer, a simplified dicing-free micromachining process has been developed for a mass-production with a high yield. The backside enhancement structures under the micro-mirror help improve the flatness of the mirror surface. A 2D Lissajous pattern scanning strategy [32,33] has been used by actuating the two axes of the scanner in resonant modes with driving frequencies in a tunable range. We have realized 2D imaging with a frame rate of up to 5 Hz, which is sufficient for clinical applications. In the following sessions, we will describe the 2D resonant MEMS scanner and the miniaturized MEMS-based intraoperative confocal microscope.

2. 2D Au-Coated Resonant MEMS Scanner Development

2.1. Design of the 2D Au-Coated Resonant MEMS Scanner

The ray-tracing simulation has been studied in the optics design software (ZEMAX, ver. 13, Kirkland, WA, USA) to design the optics and optomechanical system for the scan-head (OD 5.5-mm package) of the miniature fiber-based NIR fluorescence intraoperative confocal microscope, as shown in Figure 1. A 2D MEMS scanner was located at the post-objective position that was close to the distal-end of the parabolic mirror (focus: 4.6, OD 5.0 mm, Al substrate, custom-made by diamond turning).

Single-mode fibers (model: S630HP, optimized for the NIR range, NA = 0.12, Nufern, East Granby, CT, USA) have been used for delivering the illumination beam and collecting the fluorescence beam ($1/e^2$ diameter ~900 μm); see Figure 1a. More details about the imaging system will be introduced in Section 3.2, including the fiber-based multi-color laser system and the multi-channel fluorescence collection system (mounted on an imaging cart). Although the confocal microscope was designed for multi-color imaging, we only focused on the single-color NIR fluorescence imaging (excitation: 785 nm, emission: >800 nm) in this study. Two collimated and parallel beams (excitation and emission) were weakly focused by the front-side parabolic mirror before being reflected by the MEMS scanner; see Figure 1a. At the center of the parabolic mirror, a solid immersion lens (SIL, fused silica, OD 1.9 mm, full hemisphere, n = 1.46, radius: 1.5 mm) contacted the tissue specimens. The distance between the two collimated and parallel beams was 3.8 mm, which were precisely aligned by a pair of Risley prisms (BK7 glass, OD 1 mm, 1 mm thick, wedge angle: 0.1 deg. ±1.5 arcminutes) are shown in Section 3.1. The MEMS scanner steered the light beams with large tilting angles, around the Y-axis and the X-axis (Figure 1b,c) and achieved a large field-of-view (FOV). Based on a custom-made spring-based mechanism, a Z-axis piezoelectrical actuator (P-601.4SL, Physik Instrumente, Karlsruhe, Germany) performed the axial scanning with either a DC stacking mode or a slow scanning mode (<5 Hz); details in Section 3.1. An FOV of 800 μm on the lateral axis can be achieved by scanning the micro-mirror with a ±3° mechanical scanning angle (MSA).

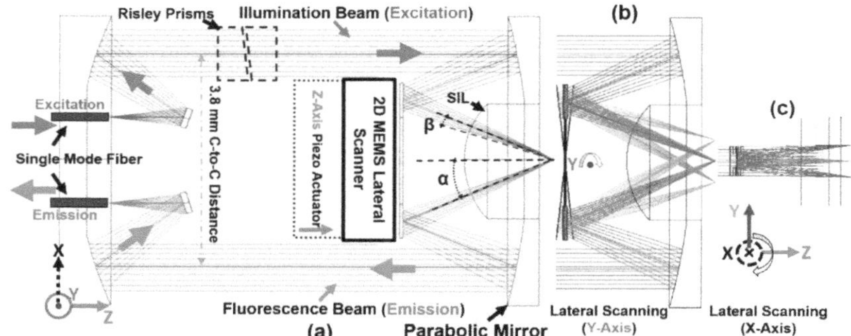

Figure 1. Ray-tracing simulation of the optical design for the MEMS-based intraoperative near-infrared (NIR) confocal microscope's scan-head (outer diameter (OD): 5.5-mm package). (**a**) Schematic drawing of the scan-head, such as the collimating, focusing, and scanning in the post-objective dual-axis confocal architecture aimed for 3D NIR fluorescence imaging, demonstrating the geometric requirements for the MEMS scanner; a pair of tiny Risley prisms were used for precise alignment; single-mode fibers (S630HP, numerical aperture (NA) = 0.12) were used for delivering and collecting light beams; SIL: solid immersion lens, β: free-space numerical aperture of the individual beams, and α: the intersection half-angle of the beams. (**b**) Lateral scanning around the Y-axis of the micro-mirror. (**c**) Lateral scanning around the X-axis of the micro-mirror.

We have used both the ray-tracing simulation and theoretical equations to design and optimize the optics. Both lateral and axial resolutions (full width at half maximum (FWHM), theoretical) may also be calculated using the following Equations (1) and (2), assuming Gaussian beams:

$$\Delta\text{Res\_lateral} = \frac{0.466\lambda}{n\beta\cos(\alpha)}, \quad (1)$$

$$\Delta\text{Res\_axial} = \frac{0.466\lambda}{n\beta\sin(\alpha)}, \quad (2)$$

where ΔRes_lateral is the lateral resolution, ΔRes_axial is the axial resolution, n is the index of refraction (assuming $n = 1.4$), β is the free-space numerical aperture (NA) of the individual beams (excitation and emission), and α is the intersection half-angle of the beams. β and α have been illustrated in Figure 1a. $\lambda = 0.785$ μm (NIR light), $\beta = 0.128$, $\theta = 0.419$ rad (24°). We have calculated the theoretical resolutions of the optics: ΔRes_lateral = 2.24 μm, ΔRes_axial = 5.02 μm. Based on the theoretical values, the confocal microscope may potentially provide cellular-resolution imaging.

For clinical applications, miniaturized intraoperative NIR confocal microscopies usually require high sensitivities for the emitted weak fluorescence signals, a large FOV with a low driving voltage (safety voltage 36 V), and compact form factors. Therefore, a parametrically-resonant 2D MEMS scanner with a patterned Au-coated surface (>90% reflectivity for an NIR > 785-nm wavelength) will be an ideal choice for the scan engine inside the confocal microscopes. In this project, the proposed gimbal frame-based 2D MEMS scanner was an advanced design based on our previous 1D parametrically-resonant scanner design [34,35] for miniaturized confocal microscopes [29]. As shown in Figure 2a,b, the geometric requirements for the 2D MEMS scanner have been determined in the ray-tracing simulation and the CAD drawing of the distal end of the fiber-based confocal microscope's scan-head (OD 5.5 mm). The actual beam width ($1/e^2$) on the micro-mirror changed (Figure 2b) due to the axial scanning along the Z-axis from 0–400 μm. An effective "dumbbell shaped" reflective area of 680 by 2900 μm^2 (Figure 2b) will be sufficient for steering the excitation and emission beams. The 2D MEMS scanner will be micro-machined using a single SOI wafer (40 μm device silicon/2 μm buried oxide/500 μm handle silicon). It used an in-plane comb-drive actuator configuration, as shown in Figure 2c–e, based on the parametric resonance working principle.

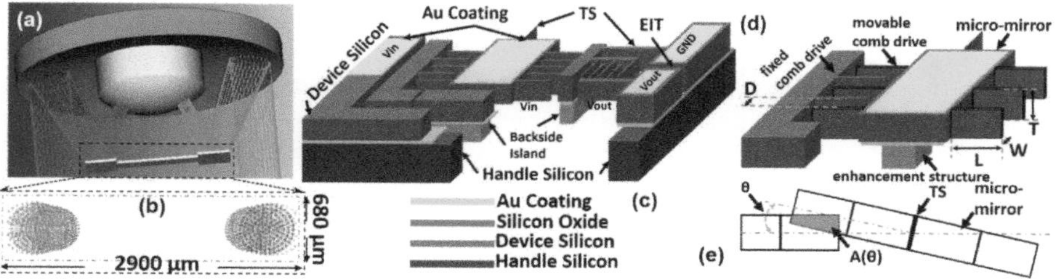

Figure 2. Schematic drawing of the 2D patterned Au-coated resonant micro-scanner and its electrical layout. (**a**) CAD drawing of the scanning micro-mirror inside the confocal microscope's scan-head (OD: 5.0 mm). (**b**) Zoom-in view of the ray-tracing beam spots (changes due to axial scanning, 0–400 μm), effective area of the "dumbbell" shaped micro-mirror. (**c**) Schematic drawing of the device silicon/buried oxide/handle silicon layers (not to scale), in-plane comb-drive actuator fingers, Au coating, inner and outer torsion springs, electrical insulation trenches, gimbal frame, backside island, and electrical layout, not to scale Note: EIT, electrical insulation trench; TS, torsion spring. (**d**) Schematic drawing of the fixed and movable comb-drive actuator fingers, D: the distance (or gap) between the comb-drive actuator fingers, W: width, L: length. (**e**) Schematic drawing of the actuation using the in-plane comb-drive actuator fingers, θ: tilting angle, A(θ): overlapped area.

For each axis of the 2D parametrically-resonant MEMS scanner, the equation of dynamic motion is essentially governed by the theory from the seminal work on parametric resonance [30]:

$$J\ddot{\theta} + c\dot{\theta} + k\theta = F(t, \theta) \quad (3)$$

where θ is the tilting angle, as shown in Figure 2e, J is the mass moment of the inertia, k is the torsion spring stiffness constant, c is the average damping constant, and F is the applied torque. The parameters were used for either the outer gimbal frame or the inner micro-mirror.

The applied torque F and the capacitance C are defined as the following Equations (4) and (5):

$$F = N\frac{1}{2}\frac{dC}{d\theta}V^2(t), \tag{4}$$

$$C = \frac{\varepsilon_0\varepsilon_r A(\theta)}{D}, \tag{5}$$

where N is the number of comb-drive fingers on one actuation side, C is the capacitance between comb-drive fingers, $dC/d\theta$ is the rate of change of the capacitance for one comb-drive finger with respect to the angular displacement, $V(t)$ is the driving signal (a periodic square waveform with a 50% duty cycle was used in our study), ε_0 is the electric constant (8.8542×10^{-12} F m^{-1}), ε_r is the relative static permittivity (1.0 for the ambient air), $A(\theta)$ is the overlapped area of the electrodes and two comb-drive actuator fingers, and D is the distance (or gap) between the electrodes and two comb-drive fingers.

Equations (3)–(5) have been used to guide the 2D resonant MEMS scanner design for both the gimbal frame (outer axis) and the micro-mirror (inner axis), assuming there is no cross-talk between these two axes. According to Equations (4) and (5), to maintain large applied torques F, low driving voltages can be achieved by increasing the number (N) and the capacitance (C) of the comb-drive fingers. While we designed the torsional springs for both axes, the tilting (or torsional) modes have to be the dominant vibration modes, and the modal analysis in ANSYS has confirmed that the eigenfrequencies (natural frequencies) of other modes were well separated from the basic tilting (torsional) mode frequencies.

In Table 1, we have listed the details about the structures. These features have been chosen mainly based on the three factors: (1) to meet the requirements in the system-level optics design and electrical layouts; (2) to consider the realistic capabilities of the microfabrication tools; (3) empirical experience. For example, the gimbal frame-based 2D resonant MEMS scanner design used electrical insulation trenches (EIT, 5 μm gap) for dividing the electrical layouts on the inner and outer axes. The width (W) and the gap (D) for the individual comb-drive fingers were designed to be 5 μm based on the capabilities (like aspect ratio) of the deep reactive ion etching (DRIE) tool. For the outer gimbal frame, four banks of comb-drive fingers have been designed on each side (symmetric on the chip); see Figure 3. In each bank, there were $N = 22$ comb-drive fingers. For the inner micro-mirror, there were $N = 115$ comb-drive fingers on each side (symmetric). The length (L), gap (D), and number (N) of comb-drive fingers have been chosen to ensure sufficiently large driving torque at a low driving voltage (~36 V) and to avoid the lateral pull-in effects. The comb-drive fingers, the torsion springs, the gimbal frame, and the scanning micro-mirror were all on the device silicon layer (40 μm thick) of the SOI wafer. Without filling in the trenches with silicon dioxide or silicon nitride [31], the gimbal frame was held by a backside island layer (around 50 μm thick) formed by the backside step-etching process on the handle silicon layer (500 μm thick) during the microfabrication process. As shown in Figure 3, inside the gimbal frame, on the four-bar inner piers, eight evenly-distributed inner torsional springs minimized the interruption of the light beam on the effective area and potentially reduced the overall stress during dynamic scanning. By taking advantage of the backside step-etching process (the same process for the backside island under the gimbal frame), the enhancement structures (~50 μm thick) will also be micro-machined under the scanning micro-mirror; see Figure 2d. The overall optical quality of the micro-mirror benefited significantly from both enhancement structures and the high-quality patterned Au/Ti coating (Ti: adhesion layer). Instead of using Cr, the Ti material has been chosen because it had less residual stress after deposition. In the previous research on various MEMS scanners, full-wafer non-patterned metallic (Al or Au) coating approaches have been commonly used [8,15]. In those processes, only a thin metallic layer (less than 500 Å) can be coated to avoid creating a short-circuit (especially on the comb-drive fingers). An alternative way for the patterned metallic coating is to use a shadow mask at the end of the microfabrication process flow. However, the shadow mask-based coating process could still potentially shorten the comb-drive fingers or encounter serious misalignment issues due to the relatively coarse alignment (compared to precise lithography). Our proposed patterned (by lithography) Au/Ti coating, at the beginning of the microfabrication process, led to a superior optical quality for the NIR fluorescence microscopy applications, because of two main factors: (1) accuracy of the reflective area (other non-effective areas will not reflect light); and (2) relatively thick (>1000 Å) metallic coating for a high reflection coefficient (>90%).

Table 1. Structural features of the 2D resonant MEMS scanner. L, length; W, width; D, distance (gap); T, thickness.

Chip Size (mm)		Comb-Drive Fingers (μm)			Torsion Springs (μm)				Gimbal Frame (mm)		Micro-Mirror (mm)		Backside Island (μm)
					Inner (8)		Outer (2)						
L	W	L	W	D	L	W	L	W	L	W	L	W	T
3.2	2.9	100	5	5	100	5	175	10	3.04	1.36	2.9	0.68	50

By using the modal analysis in ANSYS, the 2D parametrically-resonant MEMS scanner's eigenfrequency (natural frequencies) analysis has been studied; see Figure 3. Based on the parametric resonance working principle [30], the driving frequency was twice the resonant frequency over n ($f_{driving} = 2 \times f_{resonant}/n$, $n = 1, 2, 3, \ldots, N$). In ambient air at room temperature, the resonant mode can be observed with N ranging from 1–4, depending on the design of gimbal structures and torsion springs. From the modal analysis, the outer gimbal frame's tilting mode resonant frequency was around 1090 Hz (slow, around Y-axis); see Figure 3a. The inner micro-mirror's tilting mode resonant frequency was around 6250 Hz (fast, around the X-axis); see Figure 3b. The combination of the inner and outer resonant frequencies was carefully designed for the en face 2D imaging with a 2D Lissajous scan pattern (up to five frames per s). Other higher order resonant modes have also been designed (third mode: ~11,230 Hz, fourth mode: ~13,830 Hz; fifth mode: ~13,870 Hz) to ensure that they were far away from the first two basic tilting (torsional) modes.

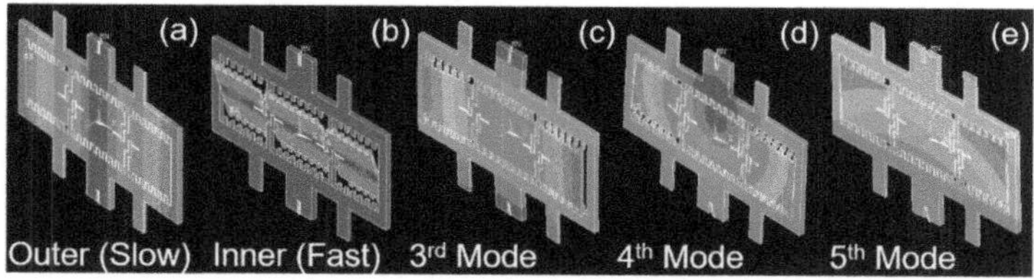

Figure 3. Finite element analysis (FEA) simulation and modal analysis in ANSYS for the 2D parametrically-resonant MEMS scanner. (**a**) Outer (slow) tilting around the Y-axis (resonant frequency: ~1090 Hz); (**b**) inner (fast) tilting around the X-axis (resonant frequency: ~6250 Hz); (**c–e**) higher order resonant modes are designed to be far away from the basic tilting modes around the X- and Y-axes, 3rd~5th mode: ~11,230 Hz, ~13,830 Hz, ~13,870 Hz, respectively.

2.2. *Microfabrication of the 2D Au-Coated Resonant MEMS Scanner*

Based on a four-inch SOI wafer, a four-mask, dicing-free microfabrication process has been developed for a high-yielding (>80%) mass-production of the 2D resonant MEMS scanners. The SOI wafer consisted of a device silicon layer (40 μm thick), buried oxide layer (BOX, 2 μm thick), and a handle silicon layer (500 μm thick). As shown in Figure 4, MEMS chips (footprint size: 3.2 by 2.9 mm^2) were dry-etched and dry-released through three primary etching steps using the DRIE process (SPTS Pegasus, fluorine-based Bosch Process; etching rate was ~6 μm/min). The process started with a high-quality silicon dioxide (SiO_2, 2 μm thick) layer deposition on both sides of the SOI wafer using the low pressure chemical vapor deposition (LPCVD) process equipment: the Tempress system for low temperature oxide (LTO)); see Figure 4a. The standard oxidation recipe for the LPCVD process is shown as follows: 425 °C temperature, 150 mTorr pressure, O_2 gas flow rate of 225 standard cubic centimeters per minutes (SCCM), N_2 gas flow rate of 100 SCCM, SiH_4 gas flow rate of 75 SCCM. Then, the backside the SiO_2 layer was patterned (Mask 1, handle Si chip frame) with the reactive ion etching (RIE) process (LAM 9400); see Figure 4b. The key parameters in the recipe of the SiO_2 etching process are listed as follows: 2.3 mTorr pressure, CHF_3 gas flow rate of 5 SCCM, bias voltage 250 V. A thin photoresist layer (SPR 220, 5 μm thick, 2500-rpm spin speed, 30-s soft-bake, step-down to 115 °C/90 s MicroChem, Westborough, MA, USA) has been used for the patterning of SiO_2. The SiO_2 layer essentially performed as a hard mask for the handle silicon layer's full-depth etching in Figure 4g. As shown in Figure 4c, another SiO_2 layer (2 μm thick) by the plasma-enhanced chemical vapor deposition (PECVD) process (GSI ULTRADEP 2000, GSI Lumonics, Novanta Inc., Bedford, MA, USA) on the handle silicon layer was patterned (Mask 2, island) for the step-etching process, to form the islands under the gimbal frame and the enhancement structures under the micro-mirror. For the SiO_2 deposition by the PECVD process, the key parameters of the deposition recipe are listed as follows: temperature 250 °C, SiH_4 gas flow rate of 100 SCCM, N_2O gas flow rate of 300 SCCM, RF power 22 W. The Au/Ti (thickness: 1000 Å/50 Å; Ti: adhesion) metallic coating layer was first prepared by an evaporation machine (Enerjet) onto the front-side device silicon layer. As shown in Figure 4d, the Au/Ti metallic coating layer was then patterned (Mask 3, Au coating) by the lift-off process to form the reflective surface on the scanning micro-mirror for the NIR light beam, electrical pads, and the alignment marks around the chip. The metallic coating layer will be fully protected through the rest of the processes so that the effects of the sequential processes on the surface roughness are minimal. The structures with fine features (up to 5 μm resolution) on the front-side device silicon layer (40 μm thick) were formed by the DRIE process (Mask 4, device Si). These important structures include a scanning micro-mirror, a gimbal frame, comb-drive actuator fingers (100-μm length, 5-μm width, 5-μm gap), outer torsion springs (two springs, each had a 10-μm width, 175-μm length), inner torsion springs (eight, each had a 5-μm width, 100-μm length), and electrical insulation trenches (5-μm gap); SEM images are shown in Figure 5. During the DRIE process, an inductively-coupled plasma (ICP) source has been used (825 W power, 2 M Hz). In the chamber, several key parameters have been controlled: 23 mTorr pressure, 40 °C coil temperature, 20 °C substrate temperature, SF_6/Ar (100/40) with bias (9 W). Before processing the backside of the SOI wafer, the front-side features had to be fully protected by spin-coating thick photoresist (AZ9260, 10 μm thick, spin speed 2000 rpm, soft-bake 110 °C/180 s, Clariant Corporation, Muttenz, Switzerland), which not only covered the surface, but also filled in the trenches on the device silicon. A thick photoresist layer (AZ9260, 8 μm thick, spin speed 3000 rpm) needed to be patterned one more time (Mask 1, handle silicon chip frame) on the backside handle Si layer; see Figure 4f. During the whole backside step-etching process (Figure 4f,g), the four-inch wafer needed to be fully attached to the six-inch carrier wafer with thermally-conductive perfluoropolyether (PFPE) oil, which would significantly enhance the thermal transfer. The wafer or photoresist burning incidents are common issues in the development of MEMS scanners. These problems usually occur due to overheating or poor thermal transfer during the dry etching of the thick handle silicon layer (500 μm) at the full wafer level, especially when the front-side device silicon layer is already full of structures, such as the trenches and comb-drive fingers. The PFPE oil helped resolve the thermal transfer problems by enhancing the contact between the carrier wafer and the SOI wafer. The step-etching process on the handle silicon layer had to be precisely time controlled by reaching the buried oxide layer; see Figure 4g. The step-etching process on the handle silicon layer would form two critical structures (around 50 μm thick): (1) the backside island for supporting the outer gimbal frame of the MEMS scanner; and (2) the enhancement structure under the micro-mirror for a high optical quality. Finally, the buried oxide layer under the moving structures on the device silicon layer (such as the scanning micro-mirror, torsion springs, and the gimbal frame) was released by a buffered oxide etch (BOE, 7:1); see Figure 4h. The conventional dicing saw-based "wet" cutting process at the last step of the microfabrication process could potentially damage the fragile micro-mirrors or gimbal frames that were linked to the substrates only by a few torsion springs. These issues can be avoided in the newly-developed dicing-free process by combining front-side and backside DRIE, leading to improved die yielding (>80%) and MEMS chips with arbitrary contours. The individual chips were dry-released from the SOI wafer by manually breaking the struts (link arms, 15 μm width) on the edge of the chip, in Figures 4h and 5a, with either the laser cutting dry process [35] or the torque applied by a sharp tweezer tip. A custom-made MEMS probe station (Model S-725PLM&-PRM, Signatone, Gilroy, CA, USA) has been used for screening the dry-released chips, which would be integrated into the distal end scan-head of the miniaturized intraoperative NIR fluorescence confocal microscope.

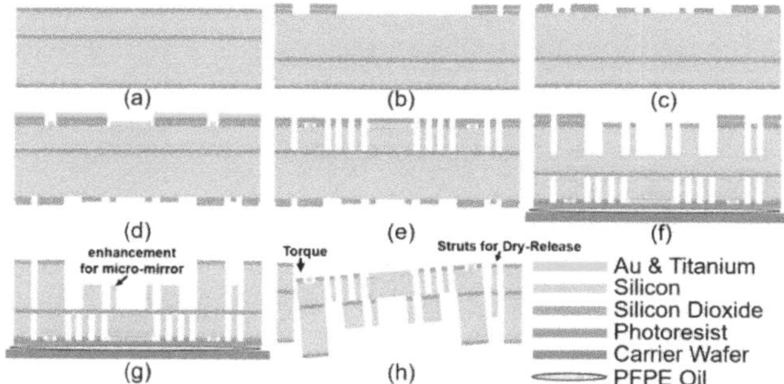

Figure 4. Dicing-free single SOI-wafer-based microfabrication process flow for the 2D patterned Au-coated resonant MEMS scanner with backside enhancement structures. (**a**) Wafer cleaning and the SiO_2 hard mask layer preparation on both sides of the SOI wafer using LPCVD; (**b**) patterning the backside SiO_2 hard mask for the scanner's full chip frame on the handle silicon layer (Mask 1, handle silicon chip frame); (**c**) patterning the backside SiO_2-based hard mask (Mask 2, island); this SiO_2 layer was prepared by PECVD; (**d**) lift-off process for a patterned Au/Ti coating (1000 Å/50 Å) to form the reflective surface for the NIR light and the alignment marks on the scanner (Mask 3, Au coating); (**e**) front-side DRIE process on the device silicon layer (Mask 4, device silicon); (**f**) DRIE process on the backside handle silicon layer; the PFPE oil was used for enhancing the thermal conductivity between the SOI and the carrier wafer; (**g**) DRIE process on the backside handle layer till reaching the buried oxide layer; (**h**) wet etching of the SiO_2 under the scanning micro-mirror using buffered hydrofluoric acid (buffered oxide etch (BOE) 7:1); the torque was applied using a sharp tweezer tip to break the struts so that the individual MEMS chip (3.2 by 2.9 mm^2) would be dry-released and harvested with a high yield (over 80%).

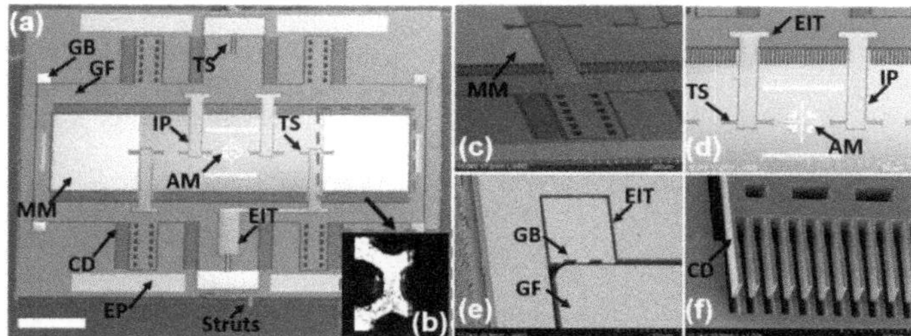

Figure 5. SEM and stereomicroscopic images of the 2D Au-coated resonant MEMS scanner. (**a**) MEMS chip with a 3.2 mm (horizontal) × 2.9 mm (vertical) footprint, scale bar: 500 μm; (**b**) zoom-in view stereomicroscopic image of the backside enhancement structures (50 μm thick) under the Au-coated scanning micro-mirror; the image was taken when the chip was upside down; (**c**) gimbal frame with the inner micro-mirror; (**d**) inner micro-mirror with "cross" shaped alignment marks at the center and two paralleled rulers; (**e**) edge of the gimbal frame with its protection bumper for the sidewall stop to avoid shorting the electricity; (**f**) comb-drive actuator fingers on the outer gimbal frame; notes: the gimbal frame was tilted with pre-loaded torque to expose the sidewall of comb-drive actuator fingers during SEM. Notes: GB, gimbal frame's bumper; GF, gimbal frame; AM, alignment marks; MM, micro-mirror coated with Au/Ti; TS, torsion spring; IP, inner piers for micro-mirror; EP, electrical pads; CD, comb-drive actuator fingers.

Scanning electron micrograph (SEM) and stereomicroscopic images of the micro-machined device are shown in Figure 5. The backside enhancement structures under the micro-mirror (Figure 5b) have improved the flatness of the micro-mirror and compensated for the residual stress induced by the Au/Ti coating (note: the image was taken when the chip was upside down). The struts (link arms, 15 μm) on the edge of the MEMS scanner were designed for the dry-releasing from the SOI wafer and can be easily broken manually by the sharp tip of a tweezer, and the residual arms are shown in Figure 5a. Different structures with fine features are shown in Figure 5c–f, including the micro-mirror, torsion springs, electrical insulation trenches, alignment marks, electrical pads, bumpers, the gimbal frame, and comb-drive fingers. The electrical pads, the coating on the micro-mirror, and the alignment marks were all formed at the same time by the Au/Ti coating and the lift-off process; see Figure 4d.

2.3. *Characterization of the 2D Au-Coated Resonant MEMS Scanner*

On the scanning micro-mirror surface, the patterned and relatively thick Au/Ti coating (1000 Å/50 Å) with a low residual stress guaranteed the high reflectivity (>90%) in the near infrared (NIR) range with low scattering losses [24,35]. The confocal microscopy (LEXT, Olympus, Tokyo, Japan) characterized the surface of the micro-mirror (at the static status) after microfabrication. Due to the backside enhancement structures, the micro-mirror of the MEMS scanner had a radius of more than 1800 mm with a peak-to-valley surface deformation <0.1 μm. The large radius (>1.8 m) of the micro-mirror's curvature proved that the stress of the micro-mirror was relatively low. Otherwise, the micro-mirror would either bend (small curvature radius) or even twist, which would induce serious misalignment or non-focus problems in the optics. These measurement results showed that the micro-mirror had a surface roughness of <5 nm. These characteristics will ensure high-quality imaging.

To demonstrate the scanning performance, the MEMS scanner was bonded onto a printed circuit board (PCB) mounted on a custom-made polymer holder with an outer diameter of 10 mm; see Figure 6a,b. The polymer holder was clamped onto a polymer-based fixture, including a V-groove and a cube. Most of the polymer-based fixtures were custom-made by a 3D printing tool (SLA, 3D Systems). Using a He-Ne laser, the scanning pattern can be observed directly on the paper screen; see Figure 6b. In addition, the dynamic characteristics and the parametrically resonant frequency responses have been characterized by using a He-Ne laser-based steering beam measurement setup; see Figure 6c. The laser beam from the He-Ne Laser (633-nm wavelength) was normally incident on the scanning mirror surface of the 2D resonant MEMS scanner and then reflected onto a position sensing detector (PSD, OnTrak, Irvine, CA, USA) that was in front of a beam splitter (50:50). A single data acquisition card (DAQ, PCI-6115, National Instrument, Austin, TX, USA) was utilized for sending out analog driving signal outputs (two-channel 12-bit analog outputs, 2.5 mega-samples per second—MS/s dual channel) and acquiring position sensing signals (two out of four-channel high-speed analog inputs, 10 MS/s per channel). The outer and inner axes of the MEMS scanner were driven by a two-channel high-voltage amplifier (gain = 20, Model 2350, Tegam), which was connected to the DAQ card by BNC cables. Figure 7 shows the response curves' characterization of both the inner and outer axes with various driving voltages. The resonant MEMS scanner was driven by a square waveform of various driving voltages (V peak-to-peak ranged from 30–80 V) with the frequency sweeping (from 1 Hz–13k Hz, up and down sweep) at the pulse width modulation (PWM) duty cycle of 50% ($f_{driving} = 2 \times f_{resonant}/n$, $n = 1, 2, 3, 4, \ldots, N$). In the ambient air at room temperature, the parametric resonance phenomena can be easily observed from N = 1–4 with driving voltages lower than 60 V. For the inner (fast) axis, the mechanical scanning angle (MSA) ±4° can be achieved while the driving frequency ($f_{driving}$) was around 11.3 k Hz (resonant frequency $f_{resonant}$ = 5659 Hz, $N = 1$, at 36 V). For the outer (slow) axis, the scanner performed tilting with MSA = ±4°, while $f_{driving}$ = 2.1 k Hz ($f_{resonant}$ = 1050 k Hz, $N = 1$, 36 V). Compared to the simulated results from the modal analysis in ANSYS, the resonant frequency's real value of both the inner and outer axes was slightly lower due to the manufacturing tolerances and slight over-etching effects on the torsional springs and backside island structures. The phase delay information can be derived by comparing the driving signals (square waveform) from

PCI-6115 DAQ (National. Instruments, Austin, TX, USA) with the position sensing signals (sinusoid waveform) acquired by the synchronized channels on the DAQ card. The phase delay values on both inner and outer axes will be critical for the confocal microscopic imaging system, including the open-loop control, the synchronization for data acquisition, and the real-time image reconstruction.

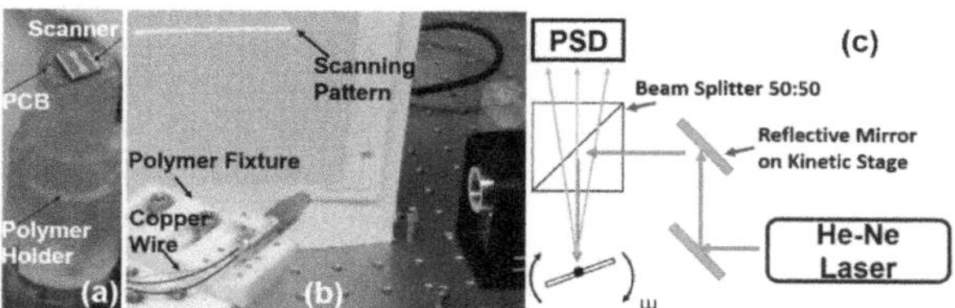

Figure 6. Characterization of the 2D Au-coated resonant MEMS scanner using a position sensing detector (PSD)-based measurement system setup. (**a**) Simple packaging of the MEMS device onto the PCB stage mounted on a 3D-printed polymer holder; (**b**) scanning pattern demonstration by clamping the holder on the 3D-printed polymer fixture; thin protected copper wires were used for delivering electricity to the MEMS scanner; (**c**) schematic drawing of the custom-made measurement system setup for fully characterizing the 2D resonant MEMS scanner; the PSD was located behind the 50:50 beam splitter; two silver-coated reflective mirrors held on kinetic stages were used to align the He-Ne laser beam accurately onto the surface of the scanning micro-mirror; the steered laser beam passed through the beam splitter and was detected by the PSD with fast responses.

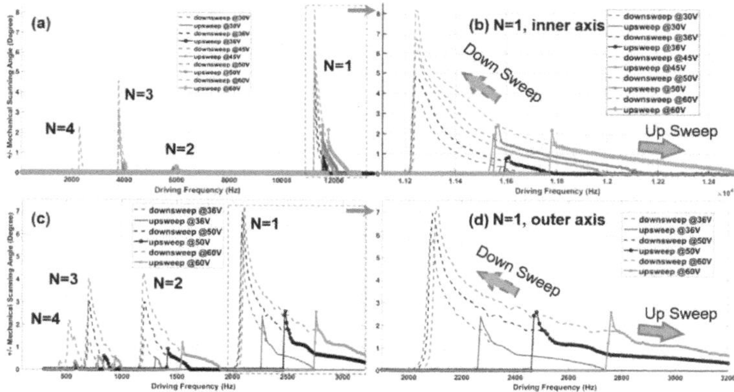

Figure 7. Characterization of the parametrically-resonant scanner on both inner (fast) and outer (slow) axes. (**a**) Tilting amplitude (mechanical scanning angle (MSA), units: degrees) vs. driving frequency sweeping (from 1 Hz–13k Hz, in ambient air, with variant driving voltage from 30 V–60 V) of the inner (fast) axis (N = 1–4); (**b**) zoom-in view of the response curves of the inner (fast) axis while N = 1; (**c**) tilting amplitude vs. driving frequency sweeping (from 1 Hz–3500 Hz) of the outer (slow) axis (N = 1–4, in the ambient air at room temperature, with variant driving voltage from 36 V–60 V); (**d**) zoom-in view of the response curves of the outer axis for N = 1. Note: while N = 1, the driving frequency is double the resonant scan frequency.

3. System Development of the MEMS-Based Near-Infrared (NIR) Intraoperative Confocal Microscope

3.1. 2D Resonant MEMS Scanner Integration in the Intraoperative Confocal Microscope

The 2D resonant MEMS scanner required precise alignment based on the on-chip alignment marks under the stereomicroscope (SMZ660 ESD Microscope, Nikon, Tokyo, Japan). As shown in Figure 8, the new 2D resonant MEMS scanner has been integrated into the distal scan-head of the miniaturized intraoperative confocal microscope. The MEMS scanner has been wire-bonded manually by a wedge wire bonder (Model 7476, Westbond, Anaheim, CA, USA) using gold wires; see Figure 8e. Additional conductive silver epoxy will be deposited onto the bonding pads to ensure the steady electrical connections aimed for longitudinal studies; see Figure 8d,e. As shown in Figure 8a, the distal end of the fiber-based confocal microscope prototype has an outer diameter of 5.5 mm, three thin protected copper wires delivered electricity to the MEMS scanner (V_{in}, V_{outer}, Ground- GND). The piezoelectrical actuated spring mechanism drove the MEMS scanner holder forward and backward (at the distal end of scan-head, dashed line), in Figure 8a,c. Two tiny Risley prisms (OD 1.0 mm) were used to align the dual beams precisely to be parallel; see Figure 8b. Additional metal fixtures helped the alignment and will be disassembled while the system integration is fully accomplished; see Figure 8b. The Z-axis piezoelectrical actuator (P-601.4SL, Physik Instrumente, Karlsruhe, Germany) was used for the axial scanning (DC stacking mode and slow Z-axis scanning with 5 Hz or less); see Figure 8c. Figure 8d,e essentially demonstrates the axial scanning of the MEMS scanner holder driven by the Z-axis piezoelectrical actuator (P-601.4SL) through the spring mechanism (350-μm range, up to 5 Hz, sufficient for vertical cross-sectional imaging with a raster scanning pattern).

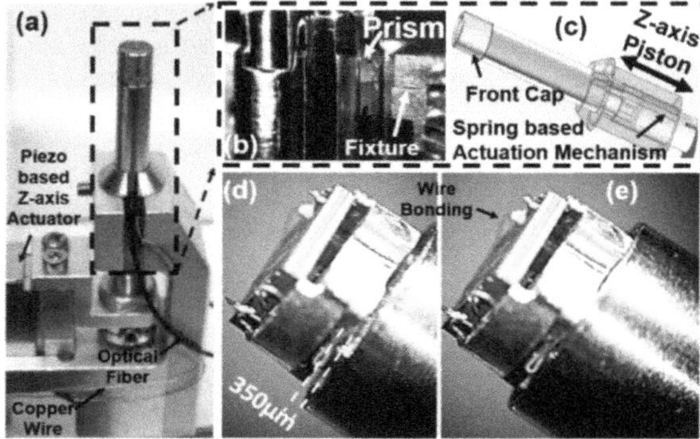

Figure 8. Integration of the 2D resonant MEMS scanner in the miniaturized fiber-based intraoperative NIR fluorescence confocal microscope. (**a**) Stereomicroscopic image of the microscope; a piezoelectrical Z-axis actuator (P-601.4SL) was used for axial scanning (DC stacking mode, or axial scanning up to 5 Hz with a 350-μm range); (**b**) zoom-in view of the microscope's distal end; Risley prisms were used for aligning the beams precisely to be parallel; additional metal alignment fixtures; (**c**) schematic drawing of the scan-head's distal end and the spring-based Z-axis piston mode actuation mechanism; (**d**,**e**) zoom-in view of the Z-axis movement of the MEMS scanner holder stage with 350-μm movement; the front cap has been taken off.

3.2. Fiber-Based Fluorescence Imaging System Development

A single DAQ card (PCI-6115, deep onboard memory = 32 MS) has been utilized for the NIR fluorescence intraoperative confocal microscopic imaging system. A custom-made NI-DAQmx-based LabVIEW (National Instrument) program has been developed for controlling the 2D resonant MEMS scanner and the piezoelectrical actuator (P-601.4SL) with feedback controls, acquiring fluorescence signals, and reconstructing 2D images.

As shown in Figure 9, a multi-color fiber-coupled laser engine has been developed for the intraoperative confocal microscope system (excitation with 660-nm and 785-nm CW lasers from Toptica company, Munich, Germany; more lasers can be added inside; only a 785-nm CW laser was used in this study). The fiber-based multi-channel fluorescence collection system has been developed for splitting the fluorescence emission signals acquired from the confocal microscope, in Figure 9. The fluorescence emission light through the single-mode fiber (FC/PC connector) would be first collimated by an aspheric lens (C240TME-B, Thorlabs, Newton, NJ, USA) and then collected by the photomultiplier tubes (PMT,

H7422P-50, Hamamatsu, Japan) through a dichroic mirror (cut-off wavelength: 750 nm), long-pass filters, and condense lenses (C240TME-B, Thorlabs). The dichroic mirror split the fluorescence emission lights into two different channels (<750-nm light was reflected, >750-nm passed through). In this study, only the NIR (>800 nm, excited by a 785-nm laser) fluorescence collection channel was used. I-V low noise amplifiers (FEMTO, Model DHPCA-100, no bias, V/A = 10^5, low-pass 10M Hz, DC coupling) band-pass filter and converted the weak current signal outputs from the PMT detectors, which detected the emitted fluorescence light from the confocal microscope. On the DAQ card, high-speed analog input (10 MS/s per channel) channels digitized the analog signals from the low-noise I-V amplifier, while analog output channels generated the control signal of the 2D MEMS resonant scanner at 1 M samples/s. For fluorescence 2D imaging (up to five frames/s, 2D frame rate), amplified square waveforms (driving voltage Vpp = 36 V) at 11.3 k Hz and 2.1 k Hz drove the inner axis and the outer axis, respectively. Both the X- and Y-axes of the micro-mirror operated in open-loop resonant scanning modes. A 2D Lissajous scan pattern [32,36] has been carefully designed with an optimized frequency combination (validated in a custom-made MATLAB program, but will not be discussed here). For en face 2D imaging on the XY-plane, in the custom-made LabVIEW program, control signals and data acquisition have been fully synchronized for real-time reconstruction of the images by remapping the time series signals to a 2D XY-plane image with a pre-calculated look-up table (LUT) for the 2D Lissajous scan pattern.

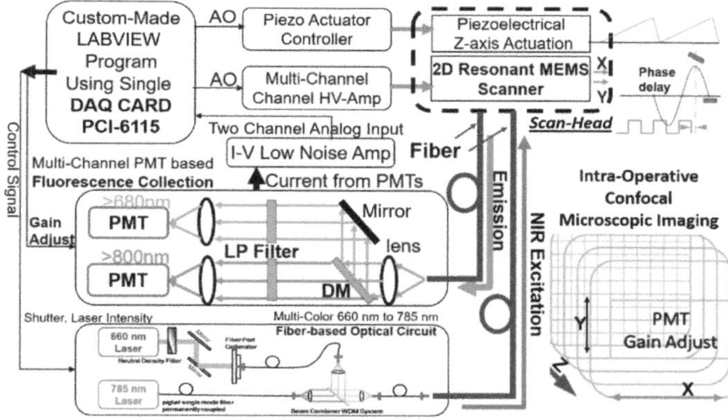

Figure 9. Schematic drawing of the NIR fluorescence confocal imaging system, including the fiber-based multi-color laser excitation and multi-channel fluorescence emission collection system based on photomultiplier tubes (PMT), control system, signal amplification, data acquisition, and real-time image reconstruction; the gain settings of the PMTs were adjusted along the Z-axis for depth imaging. Note: LP filter: long-pass filter, DM: dichroic mirror.

3.3. NIR Fluorescence Imaging Results of the Intraoperative Confocal Microscope

To characterize the NIR fluorescence confocal imaging performance of the 2D resonant MEMS-based intraoperative confocal microscope, lateral and axial resolutions (on-axis) were tested using a knife-edge target at the NIR excitation (785 nm) in the reflective mode. The tested lateral and axial resolutions were 3.5 μm and 6.5 μm, respectively, which were less than the theoretical ones (2.24 μm and 5.02 μm) due to non-perfect optical alignments. Ex vivo fluorescence imaging has also been characterized with human colon tissue specimens that were topically stained with the IRDye800CW dye (excitation: 785 nm), with an approved protocol for human subject research at Stanford University. In this study, we only used an NIR laser (785 nm, 40 mW) and the NIR fluorescence collection channel (>800 nm). The laser power (out of the SIL) into the tissue was ~2 mW. As shown in Figure 10c,d, the MEMS-based intraoperative confocal microscope performed "histology-like" imaging with a large field-of-view (up to 1000 μm). The en face horizontal cross-sectional image (XY-plane) (Figure 10e) was acquired at a Z-axis depth = 150 μm in the tissue, while the vertical cross-sectional image shows the XZ-plane with a Z-axis depth of 350 μm. In both horizontal and vertical cross-sectional images (Figure 10c,d), crypts, colonocytes, and lumen have been visualized with cellular resolutions. The ex vivo imaging experiments essentially have validated the feasibility of the intraoperative confocal microscope system.

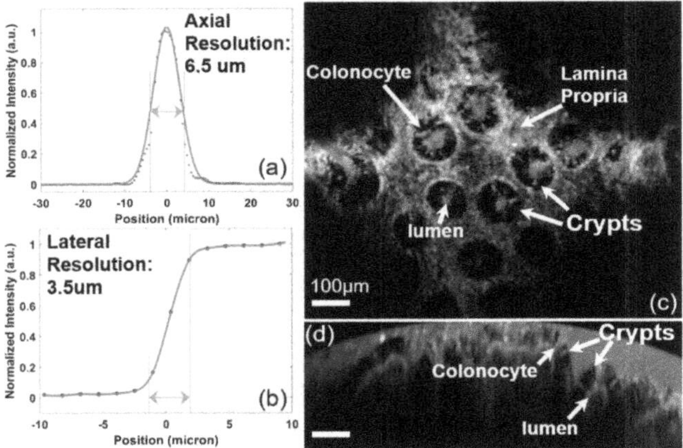

Figure 10. Fluorescence imaging results of the NIR fluorescence intraoperative confocal microscope. (**a**) Lateral and (**b**) axial resolution at the wavelength of 785 nm (focus at 150 μm out of the SIL) by measuring the FWHM in the reflective mode; ex vivo fluorescence images of human colon tissue specimens demonstrated "histology-like" imaging with a large field-of-view (up to 1000 μm), in both the (**c**) en face horizontal cross-sectional image at the 150-μm depth and (**d**) vertical cross-sectional image; crypts, colonocytes, and lumen have been visualized with cellular resolutions, scale bar: 100 μm.

4. Conclusions

In this project, we have demonstrated a new 2D patterned Au-coated parametrically-resonant MEMS scanner with a compact form-factor (3.2 by 2.9 mm^2) for the NIR fluorescence intraoperative confocal microscope system. A dicing-free microfabrication process with a patterned Au coating has been developed for a high-yielding mass-production of the 2D resonant MEMS scanner with an in-plane comb-drive configuration. Both X- and Y-axes of the resonant scanner can achieve large tilting angles ±4° (mechanical scanning angle, ±8° optically) at a relatively low driving voltage (36 V, safe for humans) with a relatively broad tunable driving frequency range. The resonant MEMS scanner has been successfully integrated into a miniaturized NIR fluorescence intraoperative confocal microscope with an outer diameter of 5.5-mm packaging at the distal end. To acquire the NIR fluorescence 2D images, the resonant MEMS scanner operated at the resonant modes on both axes and performed the 2D Lissajous scan pattern. We have demonstrated the ex vivo "histology-like" 2D imaging on human colon tissue specimens with up to five frames/s. The 2D resonant MEMS scanner can also be utilized for other applications, including MEMS-based microendoscopy and wide-field endoscopy.

Author Contributions: C.-Y.Y., B.L., and Z.Q. designed the project and wrote the manuscript. C.-Y.Y. and B.L. analyzed the data and imaging results. Z.Q. supervised the project. All authors have discussed the results and made comments on the manuscript writing.

Funding: The project was supported by the National Science Foundation (NSF) (grant number 1808436) and the Department of Energy (DOE) (grant number 234402).

Acknowledgments: We would like to thank Harald Schenk and Haijun Li for their tremendous help with the resonant MEMS scanner development. We also would like to thank Shai Friedland from Stanford University. We thank the support from the Lurie NanoFabrication Facility and Argonne National Laboratory.

Conflicts of Interest: The authors declare no conflict of interest.

References

1. Pogue, B.W.; Rosenthal, E.L.; Achilefu, S.; van Dam, G.M. Perspective review of what is needed for molecular-specific fluorescence-guided surgery. *J. Biomed. Opt.* **2018**, *23*, 100601. [CrossRef] [PubMed]
2. Troyan, S.L.; Kianzad, V.; Gibbs-Strauss, S.L.; Gioux, S.; Matsui, A.; Oketokoun, R.; Ngo, L.; Khamene, A.; Azar, F.; Frangioni, J.V. The FLARE™ intraoperative near-infrared fluorescence imaging system: A first-in-human clinical trial in breast cancer sentinel lymph node mapping. *Ann. Surg. Oncol.* **2009**, *16*, 2943–2952. [CrossRef] [PubMed]
3. Fuks, D.; Pierangelo, A.; Validire, P.; Lefevre, M.; Benali, A.; Trebuchet, G.; Criton, A.; Gayet, B. Intraoperative confocal laser endomicroscopy for real-time in vivo tissue characterization during surgical procedures. *Surg. Endosc.* **2019**, *33*, 1544–1552. [CrossRef]
4. Chang, T.P.; Leff, D.R.; Shousha, S.; Hadjiminas, D.J.; Ramakrishnan, R.; Hughes, M.R.; Yang, G.-Z.; Darzi, A. Imaging breast cancer morphology using probe-based confocal laser endomicroscopy: Towards a real-time intraoperative imaging tool for cavity scanning. *Breast Cancer Res. Treat.* **2015**, *153*, 299–310. [CrossRef]
5. Carrasco-Zevallos, O.M.; Viehland, C.; Keller, B.; Draelos, M.; Kuo, A.N.; Toth, C.A.; Izatt, J.A. Review of intraoperative optical coherence tomography: Technology and applications [Invited]. *Biomed. Opt. Express* **2017**, *8*, 1607–1637. [CrossRef] [PubMed]
6. Ehlers, J.P. Intraoperative optical coherence tomography: Past, present, and future. *Eye (Lond.)* **2016**, *30*, 193–201. [CrossRef] [PubMed]
7. Kantelhardt, S.R.; Kalasauskas, D.; Konig, K.; Kim, E.; Weinigel, M.; Uchugonova, A.; Giese, A. In vivo multiphoton tomography and fluorescence lifetime imaging of human brain tumor tissue. *J. Neurooncol.* **2016**, *127*, 473–482. [CrossRef]
8. Hoy, C.L.; Durr, N.J.; Chen, P.; Piyawattanametha, W.; Ra, H.; Solgaard, O.; Ben-Yakar, A. Miniaturized probe for femtosecond laser microsurgery and two-photon imaging. *Opt. Express* **2008**, *16*, 9996–10005. [CrossRef] [PubMed]
9. Sanai, N.; Snyder, L.A.; Honea, N.J.; Coons, S.W.; Eschbacher, J.M.; Smith, K.A.; Spetzler, R.F. Intraoperative confocal microscopy in the visualization of 5-aminolevulinic acid fluorescence in low-grade gliomas. *J. Neurosurg.* **2011**, *115*, 740–748. [CrossRef] [PubMed]
10. Qiu, Z.; Piyawattanametha, W. MEMS Actuators for Optical Microendoscopy. *Micromachines* **2019**, *10*. [CrossRef] [PubMed]
11. Dickensheets, D.L.; Kreitinger, S.; Peterson, G.; Heger, M.; Rajadhyaksha, M. Wide-field imaging combined with confocal microscopy using a miniature f/5 camera integrated within a high NA objective lens. *Opt. Lett.* **2017**, *42*, 1241–1244. [CrossRef] [PubMed]
12. Dickensheets, D.L.; Kino, G.S. Micromachined scanning confocal optical microscope. *Opt. Lett.* **1996**, *21*, 764–766. [CrossRef] [PubMed]
13. Piyawattanametha, W.; Ra, H.; Qiu, Z.; Friedland, S.; Liu, J.T.; Loewke, K.; Kino, G.S.; Solgaard, O.; Wang, T.D.; Mandella, M.J.; et al. In vivo near-infrared dual-axis confocal microendoscopy in the human lower gastrointestinal tract. *J. Biomed. Opt.* **2012**, *17*, 021102. [CrossRef]
14. Piyawattanametha, W.; Wang, T.D. MEMS-Based Dual Axes Confocal Microendoscopy. *IEEE J. Sel. Top. Quantum Electron.* **2010**, *16*, 804–814. [CrossRef] [PubMed]
15. Liu, J.T.; Mandella, M.J.; Loewke, N.O.; Haeberle, H.; Ra, H.; Piyawattanametha, W.; Solgaard, O.; Kino, G.S.; Contag, C.H. Micromirror-scanned dual-axis confocal microscope utilizing a gradient-index relay lens for image guidance during brain surgery. *J. Biomed. Opt.* **2010**, *15*, 026029. [CrossRef] [PubMed]
16. Wei, L.; Yin, C.; Fujita, Y.; Sanai, N.; Liu, J.T.C. Handheld line-scanned dual-axis confocal microscope with pistoned MEMS actuation for flat-field fluorescence imaging. *Opt. Lett.* **2019**, *44*, 671–674. [CrossRef] [PubMed]
17. Arrasmith, C.L.; Dickensheets, D.L.; Mahadevan-Jansen, A. MEMS-based handheld confocal microscope for in-vivo skin imaging. *Opt. Express* **2010**, *18*, 3805–3819. [CrossRef]
18. Cho, A.R.; Han, A.; Ju, S.; Jeong, H.; Park, J.-H.; Kim, I.; Bu, J.-U.; Ji, C.-H. Electromagnetic biaxial microscanner with mechanical amplification at resonance. *Opt. express* **2015**, *23*, 16792–16802. [CrossRef] [PubMed]

19. Liu, L.; Wu, L.; Zory, P.; Xie, H. Fiber-optic confocal microscope with an electrothermally-actuated, large-tunable-range microlens scanner for depth scanning. In Proceedings of the 2010 IEEE 23rd International Conference on Micro Electro Mechanical Systems (MEMS), Wanchai, Hong Kong, China, 24–28 January 2010.
20. Liu, L.; Wang, E.; Zhang, X.; Liang, W.; Li, X.; Xie, H. MEMS-based 3D confocal scanning microendoscope using MEMS scanners for both lateral and axial scan. *Sens. Actu. A Phys.* **2014**, *215*, 89–95. [CrossRef] [PubMed]
21. Seo, Y.H.; Hwang, K.; Park, H.C.; Jeong, K.H. Electrothermal MEMS fiber scanner for optical endomicroscopy. *Opt. Express* **2016**, *24*, 3903–3909. [CrossRef]
22. Ra, H.; Piyawattanametha, W.; Taguchi, Y.; Lee, D.; Mandella, M.J.; Solgaard, O. Two-Dimensional MEMS Scanner for Dual-Axes Confocal Microscopy. *J. Microelectromech. Syst.* **2007**, *16*, 969–976. [CrossRef]
23. Qiu, Z.; Pulskamp, J.S.; Lin, X.; Rhee, C.H.; Wang, T.; Polcawich, R.G.; Oldham, K. Large displacement vertical translational actuator based on piezoelectric thin films. *J. Micromech. Microeng.* **2010**, *20*, 075016. [CrossRef] [PubMed]
24. Qiu, Z.; Rhee, C.H.; Choi, J.; Wang, T.D.; Oldham, K.R. Large Stroke Vertical PZT Microactuator With High-Speed Rotational Scanning. *J. Microelectromech. Syst.* **2014**, *23*, 256–258. [CrossRef]
25. Flusberg, B.A.; Cocker, E.D.; Piyawattanametha, W.; Jung, J.C.; Cheung, E.L.; Schnitzer, M.J. Fiber-optic fluorescence imaging. *Nat. Methods* **2005**, *2*, 941–950. [CrossRef] [PubMed]
26. Piyawattanametha, W.; Barretto, R.P.; Ko, T.H.; Flusberg, B.A.; Cocker, E.D.; Ra, H.; Lee, D.; Solgaard, O.; Schnitzer, M.J. Fast-scanning two-photon fluorescence imaging based on a microelectromechanical systems two- dimensional scanning mirror. *Opt. Lett.* **2006**, *31*, 2018–2020. [CrossRef] [PubMed]
27. Piyawattanametha, W.; Cocker, E.D.; Burns, L.D.; Barretto, R.P.; Jung, J.C.; Ra, H.; Solgaard, O.; Schnitzer, M.J. In vivo brain imaging using a porTable 2.9 g two-photon microscope based on a microelectromechanical systems scanning mirror. *Opt. Lett.* **2009**, *34*, 2309–2311. [CrossRef]
28. Qiu, Z.; Piyawattanametha, W. MEMS-based medical endomicroscopes. *IEEE J. Sel. Top. Quantum Electron.* **2015**, *21*, 376–391. [CrossRef]
29. Qiu, Z. Multi-spectral Dual Axes Confocal Endomicroscope with Vertical Cross-sectional Scanning for In-vivo Targeted Imaging of Colorectal Cancer. Ph.D. Thesis, University of Michigan, Ann Arbor, MI, USA, 2014.
30. Turner, K.L.; Miller, S.A.; Hartwell, P.G.; MacDonald, N.C.; Strogatz, S.H.; Adams, S.G. Five parametric resonances in a microelectromechanical system. *Nature* **1998**, *396*, 149. [CrossRef]
31. Schenk, H.; Dürr, P.; Kunze, D.; Lakner, H.; Kück, H. A resonantly excited 2D-micro-scanning-mirror with large deflection. *Sens. Actu. A Phys.* **2001**, *89*, 104–111. [CrossRef]
32. Seo, Y.H.; Hwang, K.; Kim, H.; Jeong, K.H. Scanning MEMS Mirror for High Definition and High Frame Rate Lissajous Patterns. *Micromachines* **2019**, *10*. [CrossRef]
33. Park, H.C.; Seo, Y.H.; Jeong, K.H. Lissajous fiber scanning for forward viewing optical endomicroscopy using asymmetric stiffness modulation. *Opt. Express* **2014**, *22*, 5818–5825. [CrossRef]
34. Qiu, Z.; Khondee, S.; Duan, X.; Li, H.; Mandella, M.J.; Joshi, B.P.; Zhou, Q.; Owens, S.R.; Kurabayashi, K.; Oldham, K.R.; et al. Vertical cross-sectional imaging of colonic dysplasia in vivo with multi-spectral dual axes confocal endomicroscopy. *Gastroenterology* **2014**, *146*, 615–617. [CrossRef] [PubMed]
35. Qiu, Z.; Liu, Z.; Duan, X.; Khondee, S.; Joshi, B.; Mandella, M.J.; Oldham, K.; Kurabayashi, K.; Wang, T.D. Targeted vertical cross-sectional imaging with handheld near-infrared dual axes confocal fluorescence endomicroscope. *Biomed. Opt. Express* **2013**, *4*, 322–330. [CrossRef] [PubMed]
36. Sullivan, S.Z.; Muir, R.D.; Newman, J.A.; Carlsen, M.S.; Sreehari, S.; Doerge, C.; Begue, N.J.; Everly, R.M.; Bouman, C.A.; Simpson, G.J. High frame-rate multichannel beam-scanning microscopy based on Lissajous trajectories. *Opt. Express* **2014**, *22*, 24224–24234. [CrossRef] [PubMed]

 micromachines

Article

Input Shaping Based on an Experimental Transfer Function for an Electrostatic Microscanner in a Quasistatic Mode

Kwanghyun Kim [1], Seunghwan Moon [2], Jinhwan Kim [1], Yangkyu Park [1,†] and Jong-Hyun Lee [1,2,*]

1 School of Mechanical Engineering, Gwangju Institute of Science and Technology, Gwangju 61005, Korea; khinmf13@gist.ac.kr (K.K.); goalermen2@gist.ac.kr (J.K.); yangkyu.park1@ucalgary.ca (Y.P.)
2 WeMEMS Co., Gwangju 61005, Korea; wemems7.msh@gmail.com
* Correspondence: jonghyun@gist.ac.kr; Tel.: +82-62-715-2395
† Present address: Department of Mechanical and Manufacturing Engineering, University of Calgary, Calgary, Alberta T2N 1N4, Canada.

Received: 24 January 2019; Accepted: 26 March 2019; Published: 27 March 2019

Abstract: This paper describes an input shaping method based on an experimental transfer function to effectively obtain a desired scan output for an electrostatic microscanner driven in a quasistatic mode. This method features possible driving extended to a higher frequency, whereas the conventional control needs dynamic modeling and is still ineffective in mitigating harmonics, sub-resonances, and/or higher modes. The performance of the input shaping was experimentally evaluated in terms of the usable scan range (USR), and its application limits were examined with respect to the optical scan angle and frequency. The experimental results showed that the usable scan range is as wide as 96% for a total optical scan angle (total OSA) of up to 9° when the criterion for scan line error is 1.5%. The usable scan ranges were degraded for larger total optical scan angles because of the nonlinear electrostatic torque with respect to the driving voltage. The usable scan range was 90% or higher for most frequencies up to 160 Hz and was drastically decreased for the higher driving frequency because fewer harmonics are included in the input shaping process. Conclusively, the proposed method was experimentally confirmed to show good performance in view of its simplicity and its operable range, quantitatively compared with that of the conventional control.

Keywords: microscanner; input shaping; open-loop control; quasistatic actuation; residual oscillation; usable scan range; higher-order modes

---

1. Introduction

MEMS (microelectromechanical systems) optical scanners are widely used micro systems, each of which is capable of driving a single mirror in two axes. In particular, the electrostatic actuation approach has advantages in terms of its size, power consumption, and integration and is therefore widely applied to such fields as optical coherence tomography (OCT) [1], laser projection display [2], and light detection and ranging (LiDAR) [3]. In these imaging systems, the usable scan range (USR), one of the important performance parameters of the scanner, depends on the uniformity of the beam steering speed. Here, the USR in percentiles, defined as the ratio of the usable range that satisfies a certain criterion (for example, the maximum deviation from a desired scan trajectory), will be used as one of the performance indices in input shaping.

Figure 1a,b show the speed uniformity of the optical scan angle (OSA) for two types of waveform in the scan outputs. In terms of the USR, the triangular waveform has advantages over the sinusoidal waveform in that there is no need for post-processing to enhance the uniformity of the images [4]. Thus, a triangular scan is generally preferred to a sinusoidal scan for high-quality images.

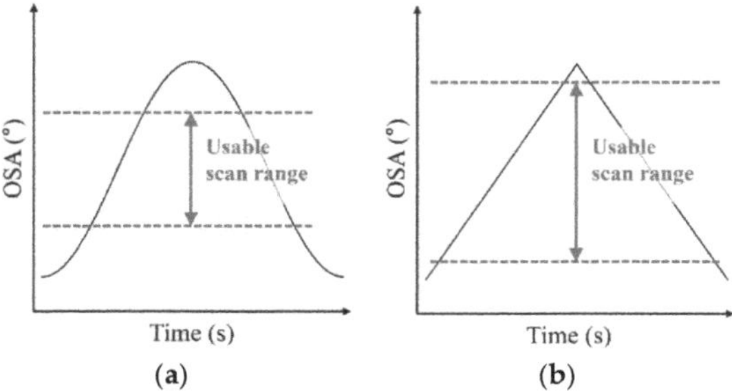

Figure 1. Definition of usable scan ranges (USRs) of the optical scan angle (OSA) for different waveforms: (**a**) sinusoidal and (**b**) triangular.

However, the input voltage in a triangular waveform may cause serious residual oscillation, resulting in distortions of the imaging [5] if one of the harmonics of the input frequency is close to or matched by the fundamental torsional mode of the microscanner. To solve this problem, studies have been performed on input shaping, such as open-loop control [6,7] and closed-loop control [8,9].

In the case of open-loop control, mechanical parameters such as the moment of inertia, damping coefficient, and torsional spring constant are extracted for a second-order dynamic model based on the measured frequency response of the microscanner. This model is used for inverse filtering to eliminate residual oscillation from a desirable arbitrary scan output [10,11].

However, the second-order transfer function as a dynamic model cannot accurately represent the spectral data, even for the fundamental mode, when the scanner has nonlinearity in the drive characteristics. As a result, distortion may appear on the waveform of the scan output, even if input shaping is performed for the given microscanner. Another disadvantage is that residual oscillation cannot be avoided if the higher modes and/or sub-resonance in the output transfer function are not negligible [12].

In the case of closed-loop control, the angular position of the mirror can be accurately adjusted by feeding back the positioning error to the controller. The angular position of the mirror can be further corrected by the additional use of the second-order transfer function which was already extracted from the open-loop control [3,9]. Consequently, closed-loop control shows good performance with high robustness regardless of the residual oscillations or external disturbance. Moreover, closed-loop control can maintain good performance despite the presence of a fatal error in the second-order dynamic modeling. However, this closed-loop control approach requires position detection devices, such as capacitive [13,14] and piezoresistive [15] sensors, making the fabrication process more complicated.

In this paper, we examine an input shaping method based on an experimental transfer function (ISETF) to effectively obtain the desired output. Unlike conventional open-loop control using a modeled transfer function, this method directly uses the experimental transfer function (ETF) with no dynamic modeling of the microscanner. The ISETF makes use of the ETF including harmonics, higher modes, and/or sub-resonance so that the transfer function can be obtained more accurately than through conventional open-loop control.

This paper is organized as follows. In Section 2, the experimental conditions, including the comb structures of the electrostatic microscanner and linearization between the driving voltage and electrostatic torque, are described. Undesired oscillation induced by the resonance frequency is also exemplified for the microscanner under testing. In Section 3, the procedure to obtain the ETF is presented. In Section 4, to quantitatively validate the method, a deviation from a desired scan output is evaluated using an electrostatic microscanner driven in a quasistatic mode. Using the variance data, the limitations of the proposed ISETF are examined with respect to OSA and frequency, and the performance of the ISETF is compared to those of other open-loop and closed-loop controls. Finally, the conclusions of this study are given in Section 5.

2. Quasistatic Microscanner

2.1. Characteristics of the Microscanner

The device used in this study is the two-axis electrostatic microscanner that was previously reported by our group [16]. As shown in Figure 2a, the fixed comb electrodes in the slow axis were tilted for quasistatic actuation, where linearization with driving voltage was conducted using a pair of comb actuators. Only the actuator in the slow axis was tested to quantitatively validate the proposed ISETF, because input shaping can be basically applied to the scanner operated in a quasistatic mode. Details of the dimensions of the electrodes are listed in Table 1.

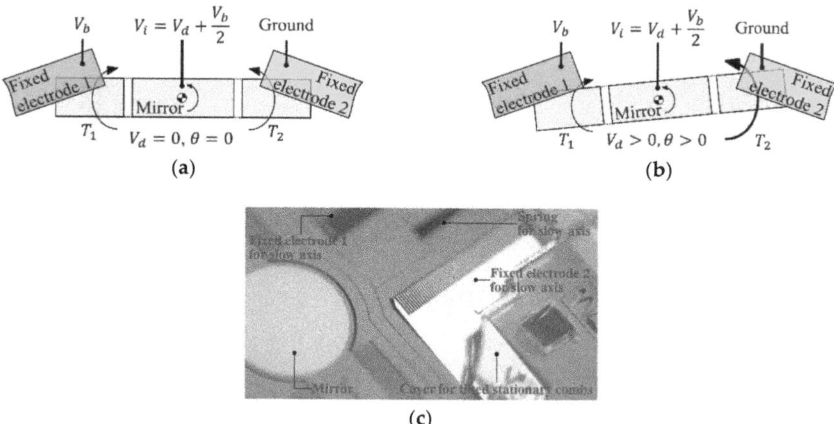

Figure 2. Linearization between electrostatic torque and driving voltage (V-linearization) for the slow axis: (**a**) initial position ($V_d = 0$), (**b**) rotated position ($V_d > 0$), and (**c**) photograph of the microscanner with tilted stationary combs.

Table 1. Dimensions of the comb electrodes in the microscanner.

Parameters	Values
Number of fingers per pair of electrodes, N	126
Thickness of the electrode, t	50 μm
Length of the electrode, l_e	175 μm
Width of the electrode, w_e	5 μm
Distance to the rotation axis, l_r	475 μm
Electrode gap, g	5 μm

2.2. Linearized Actuation

An electrostatic microscanner has the inherent nonlinearities of the electrostatic torque between two electrodes being proportional to the square of the driving voltages. The electrostatic torque can be linearized with respect to the driving voltage in the case where one pair of the electrostatic actuators is driven independently [17]. This voltage linearization (so-called V-linearization) allows the ISETF to be applicable to the microscanner because the ISETF is based on a linear transfer function.

For the V-linearization of the microscanner, the DC bias voltage (V_b), the ground, and the input voltage (V_i) are applied to fixed electrode 1, fixed electrode 2, and the mirror, respectively, as shown in Figure 2a, where V_i is the sum of the driving voltage (V_d) and $V_b/2$. The electrostatic torques (T_1, T_2) generated on movable electrodes 1 and 2 can be calculated using Equations (1) and (2), respectively. When V_d is zero, the magnitudes of T_1 and T_2 are equal to each other, staying at the initial position. Figure 2b shows that the value of V_d is greater than 0, resulting in the rotation of the mirror in the counterclockwise (positive sign) direction. Note that the total torque (T_1 + T_2) is linearly proportional to V_d, as expressed in Equation (3). Likewise, the linearization procedure can be extended to the case where the scanner is driven by alternating current (AC) driving voltage. Figure 2c shows a photograph of the microscanner with the tilted stationary combs used in this study.

The experimental OSA under the V-linearization condition is shown with respect to the direct current (DC) voltage in Figure 3, where the bias voltage, V_b, is 130 V, and the driving voltage, V_d, is in the range from −65 V to 65 V. Despite the V-linearization process, nonlinearity is still witnessed when the mirror is rotated beyond a certain critical angle. This result is believed to be mainly determined by the change rate of the overlapped area between the pair of comb electrodes.

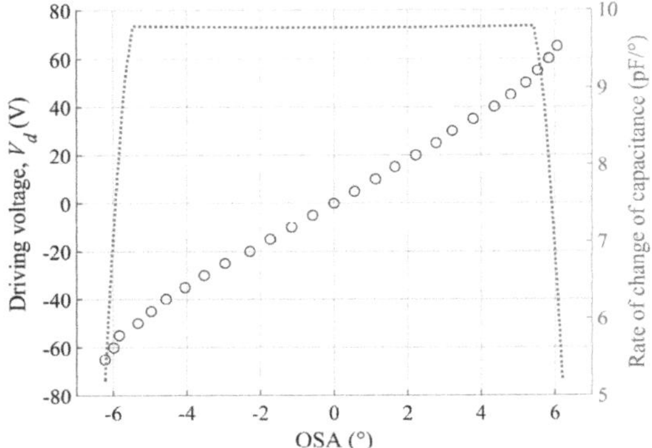

Figure 3. Nonlinear characteristics of the microscanner in quasistatic mode under the condition of V-linearization (open circles: experimental OSA; dotted line: calculated rate of change of capacitance).

Moreover, Equation (3) shows that the electrostatic torque is also proportional to the rate of change of the capacitance. To determine the cause of the nonlinearity witnessed at a large scan angle, the rate was calculated with respect to the OSA, as shown in Figure 3. The calculation result indicates that the linearity of the rate is smaller than 1% in the OSA range between −5° and 5°, whereas the rate of change of capacitance drastically decreases outside the linear range; i.e., the so-called C-nonlinearity. This supports the observation that the nonlinearity of OSA with respect to the input voltage is mainly caused by C-nonlinearity. The ISETF will be examined over the OSA range up to ±6.5° (equivalently, 13° in total OSA) to find its application limit in Section 4.1. The scan angle can be extended beyond the linear range if a new ETF is extracted at that scan angle, as long as the scan output remains periodic in a steady state.

$$T_1 = \frac{1}{2}\frac{\partial C}{\partial \theta}\left(V_d - \frac{V_b}{2}\right)^2 \tag{1}$$

$$T_2 = \frac{1}{2}\frac{\partial C}{\partial \theta}\left(V_d + \frac{V_b}{2}\right)^2 \tag{2}$$

$$T = -T_1 + T_2 = V_b V_d \frac{\partial C}{\partial \theta} \tag{3}$$

2.3. Residual Oscillation

When a triangular input voltage is provided, undesired residual oscillations can occur in the scan output. Figure 4a shows the experimental scan output with severe residual oscillations in a steady-state response when a triangular input waveform is applied at a driving frequency of 20 Hz to the microscanner. The driving frequency of 20 Hz was selected as an example, showing that this residual oscillation occurs when one of the harmonics of the input frequency might be near to the frequency of the fundamental mode in the scanner.

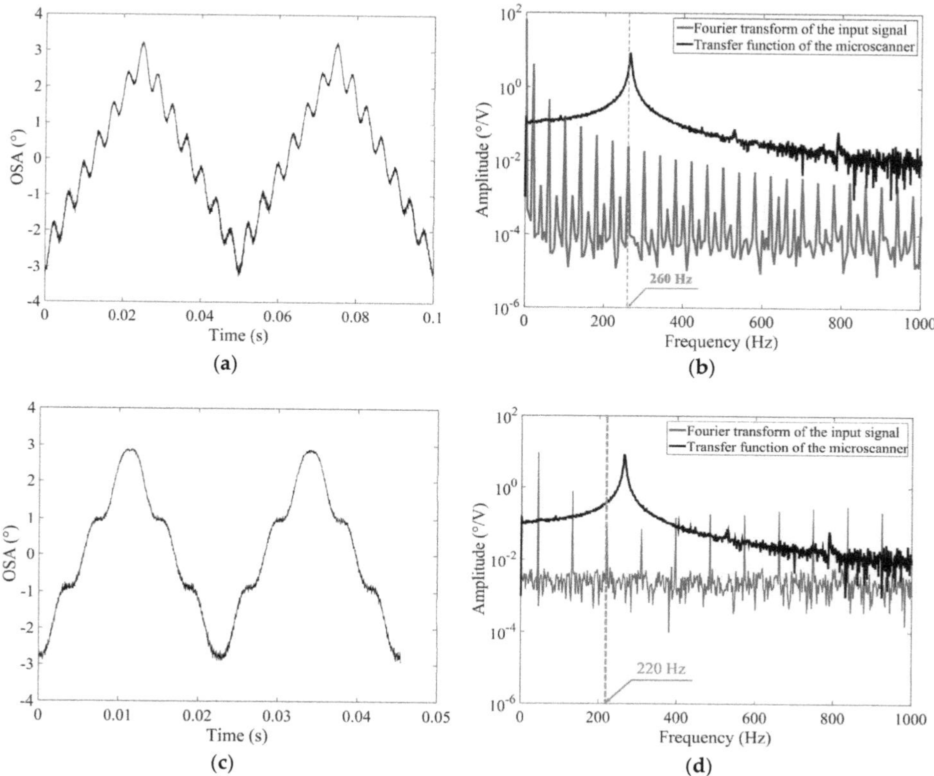

Figure 4. Exemplified dynamic response of the fabricated microscanner driven by a triangular input voltage at 20 Hz (V_d: 25 V): (**a**) scan output with severe residual oscillations in the time response and (**b**) Fourier-transformed input signal (blue) and transfer function of the microscanner (black), (**c**) scan output at 44 Hz with fewer residual oscillations, and (**d**) Fourier-transformed input signal at 44 Hz (blue) and transfer function of the microscanner (black).

Figure 4b shows the Fourier-transformed input voltage and the transfer function of the microscanner, where the sixth harmonic (260 Hz) is very close to the torsional mode at 264 Hz (the experimental procedure to obtain the transfer function will be described in detail in Section 3.1). Thus, an undesired residual oscillation of 260 Hz appears in the scan output, despite the amplitude of the sixth harmonic (260 Hz) being relatively small. Thus, the ISETF was proposed in this paper to obtain a desired output waveform by effectively reducing undesired residual oscillation. Figure 4c shows the experimental scan output at 44 Hz when the ISETF is not yet applied to the input. The residual oscillation at 44 Hz was less serious than that at 20 Hz because the harmonics of the driving frequency 44 Hz are far from the frequency of the fundamental mode, as shown in Figure 4d, compared with that of 20 Hz. The application of the proposed ISETF is not limited to these frequencies but will be evaluated for the frequency range up to 300 Hz to find its application limit in Section 4.2.

3. ISETF Procedure

3.1. Experimental Conditions

The relationship between voltage input, $X(f)$, and scan output, $Y(f)$, can be defined by a transfer function $H(f)$, as expressed in Equation (4):

$$H(f) = \frac{Y(f)}{X(f)}, \tag{4}$$

where f is the driving frequency. The ETF of the microscanner can be obtained using the impulse input or the frequency sweep. The impulse response method is advantageous in that the transfer function is simply obtained by applying a single pulse input. The pulse width should be small to include a sufficient number of high-frequency components. Meanwhile, the frequency sweep method can obtain an accurate transfer function for the high-frequency region. However, this method has several disadvantages related to being a complicated and time-consuming process because the experiment requires waiting at each measurement frequency until the steady-state response is obtained.

In this study, the impulse response method was employed to evaluate the ISETF with a pulse width of 500 μs or, equivalently, a frequency bandwidth of 2 kHz. The frequency bandwidth is large enough to include at least six frequency components for driving frequencies up to 300 Hz, over which the input shaping calculations are performed. Note that 300 Hz is higher than generally required for the scan frequency of the slow axis and is also higher than the frequency of the fundamental mode (264 Hz) of the scanner, which needs to be compensated for in most input shaping methods. The sampling rate for the ISETF was 50 kHz, which is sufficiently beyond the Nyquist criterion to provide an accurate measurement for the aforementioned driving conditions. The sampling time was set to one second, indicating that the frequency interval is 1 Hz.

It is also important that the response of the microscanner to the input pulse should remain in the linear region. Thus, in the experiment using impulse input, the magnitude of the OSA should be smaller than 5°, or, equivalently, the amplitude of the impulse voltage should be smaller than 55 V (refer to Figure 3).

3.2. Experimental Setup

The experimental setup to obtain the transfer function is shown in Figure 5, where the laser diode emits a laser beam (wavelength = 633 nm) through the collimator (Pigtailed collimator, OZ Optics, Ottawa, ON, Canada) to the mirror with an incident angle of 45°. As the first step of the experiment, the input pulse is applied to the scanner through a two-channel function generator (AFG3102, Tektronix, Beaverton, OR, USA) with a bandwidth of 240 MHz and a power amplifier (A400D, FLC electronics, Partille, Sweden) with a bandwidth of 500 kHz, consecutively. Next, a position-sensitive detector (PSD) (PSM2-45, Ontrak Photonics, Irvine, CA, USA) measures the position of the laser/beam reflected from the oscillating micromirror. Finally, the PSD signal is digitized at 50 kHz using an oscilloscope (DSO-X-4024A, Keysight, Santa Rosa, CA, USA) over a period of one second to extract 50,000 data points in total.

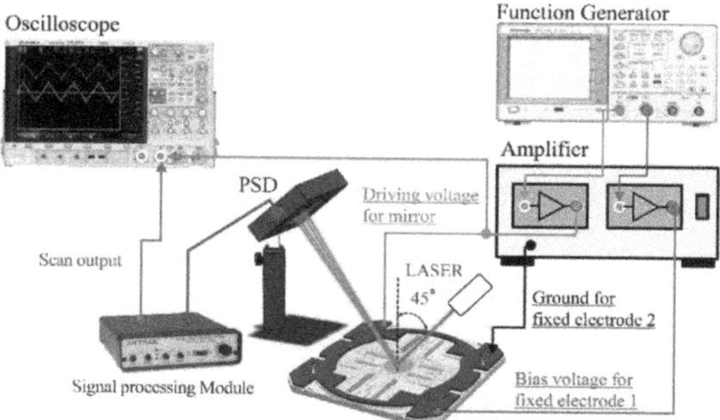

Figure 5. Experimental setup for obtaining a transfer function of the microscanner.

3.3. Extraction of the Experimental Transfer Function

Figure 6a,b represent the measured pulse input and the scan output of the mirror, respectively; these data were used to estimate the ETF through Equation (4), as shown in Figure 6c. The ETFs were averaged ten times to reduce the noise, as shown in Figure 6d. We believe that the effect of averaging on the performance of the control system lies in the effective reduction of the pulse width in the impulse response; thereby, the averaged data more clearly reveal that harmonics exist at 527 Hz and 790 Hz.

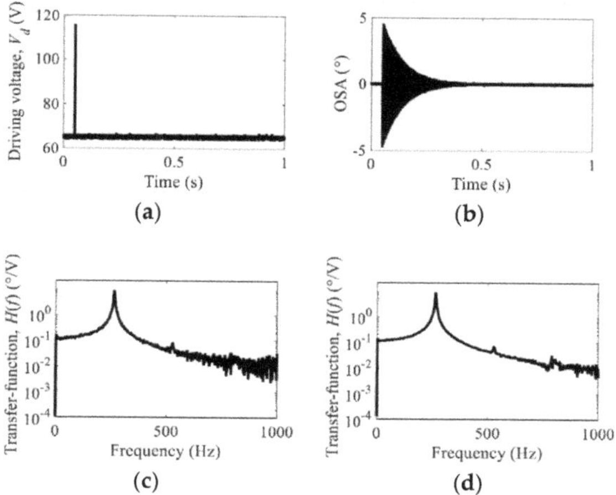

Figure 6. Estimation process of the experimental transfer function (ETF) for input shaping: (**a**) input impulse in the time domain (V_d: 55 V, $V_b/2$: 65 V, pulse width 500 μs), (**b**) output response in the time domain, (**c**) an ETF, and (**d**) a ten-times averaged ETF revealing the second and third harmonics.

Unfortunately, a conventional open-loop control that makes use of the second-order dynamic modeling can only account for the fundamental mode. On the contrary, the proposed ISETF method also has the potential to compensate for the harmonics of the fundamental mode, higher resonance modes, and sub-resonances. Nevertheless, the higher-order modes were not observed in the experiments because their frequencies exist beyond the effective bandwidth (2000 Hz), as determined by the pulse width of the input impulse. The sub-resonances, which arise from non-linear actuation forces proportional to the square of the driving voltage, were also eliminated by V-linearization, as described in Section 2.2.

3.4. Input Shaping for Triangular Output

A triangular waveform is one of the most desirable output forms, $d(t)$, in view of the USR. The ISETF method was employed to verify the effectiveness of input shaping to achieve the ideal triangular output. The shaped input signal in the time domain, $v(t)$, used to generate $d(t)$, can be calculated using the ETF obtained in Section 3.3 as follows:

(1) The ideal triangular output, $d(t)$, is Fourier-transformed to obtain $D(f)$ in the frequency domain by using Matlab software.
(2) $D(f)$ is divided by $H(f)$ to acquire the input signal in the frequency domain, $V(f)$, as shown in Equation (5).
(3) The shaped input signal in the time domain, $v(t)$, is obtained through inverse fast Fourier transform (inverse FFT).
(4) The calculated shaped input signal in the time domain, $v(t)$, is converted to the filename extension '.tfw' using ArbExpress Application software so that it can be recognized in the function generator.
(5) The converted shaped input signal is stored in the function generator through a universal serial bus (USB) memory so that it can be applied to the microscanner.

$$V(f) = \frac{D(f)}{H(f)} \tag{5}$$

To examine the performance of the ISETF described above, the input shaping was conducted for an ideal triangular output at 20 Hz with an optical scan range from $-1°$ to $1°$. Note that the scan angle is within the linear range of the microscanner, as shown in Figure 3. The shaped input signal (black line) of the ISETF produces negligible overshoots (11%) at the driving voltage, which should not put strain on the driver electronics, as shown in Figure 7a. The experimental scan output of the slow axis (black line) was directly compared with the ideal triangular output (dashed line in red) to easily quantify the deviation error (blue line), as shown in Figure 7b. The undesired residual oscillation due to the fundamental mode of the scanner was drastically diminished so that the scan output very closely approaches the ideal triangular output. Meanwhile, the model-based approach can produce significant noise in the high frequency range of the scan output. The noise was effectively removed by applying a low-pass filter, still leaving considerable residual oscillations at the harmonic frequencies of the shaped input. The scan phases were not apparently affected by the ISETF, and a quantitative evaluation of the proposed ISETF will be discussed in terms of the USR in Section 4.1.

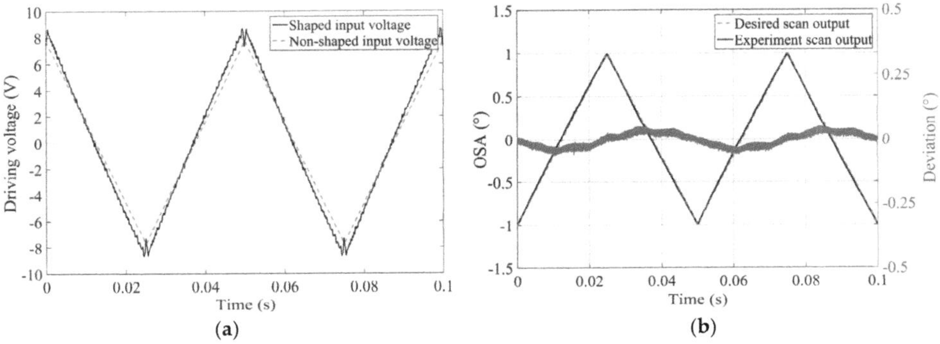

Figure 7. Performance of the input shaping method based on an experimental transfer function (ISETF): (**a**) the shaped input and non-shaped input for the optical scan range from $-1°$ to $1°$ and (**b**) comparison between the ideal triangular output and the experimental scan output obtained using the shaped input based on an experimental transfer function (ETF).

4. Application Limits of ISETF

4.1. Optical Scan Angle (OSA)

As mentioned in Section 2.2, to examine the application limits of ISETF due to C-nonlinearity, an input shaping process was experimentally implemented for total optical scan angles (total OSAs) from 2° to 12° at the driving frequency of 20 Hz. Figure 8 shows a comparison between the experimental scan output (black line) and the ideal triangular output (dashed lines in red). As the driving voltage increases, the scan output gradually differs from the ideal triangular output. More specifically, the residual oscillation is negligible up to 35 V and becomes serious at 65 V. The residual oscillation at higher angles (total OSA 13°), where the ISETF method becomes insufficient to suppress them, still contains the fundamental mode rather than other higher modes.

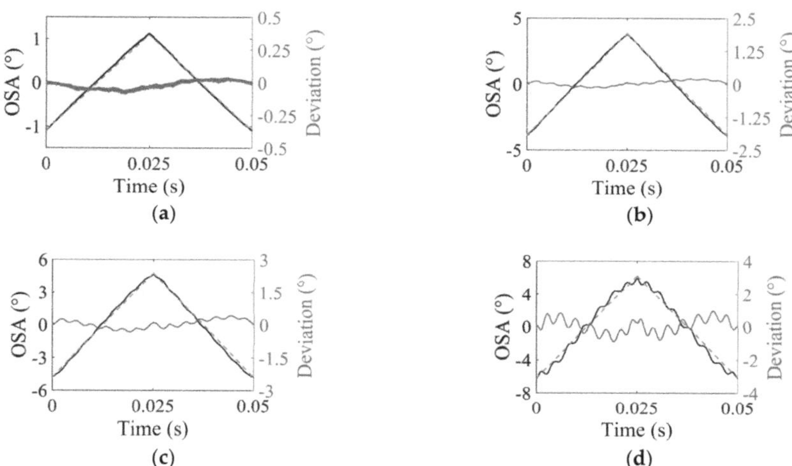

Figure 8. Performance of the ISETF with respect to the driving voltage, V_d (driving frequency: 20 Hz, sampling rate: 50 kHz): (**a**) V_d = 10 V, (**b**) V_d = 35 V, (**c**) V_d = 45 V, and (**d**) V_d = 65 V (scan output using ISETF: solid line in black; ideal triangular output: dashed lines in red).

For statistical evaluation, the root-mean-square error was calculated with reference to the ideal triangular output. Next, the calculated value was divided by the total OSA to obtain a normalized root-mean-square error (NRMSE), as a percentile, as shown in Figure 9. The NRMSE value remains at a sufficiently low level (1% or smaller) within the linear region, as confirmed in Figure 3. The NRMSE, however, shows the tendency to abruptly degrade beyond the linear region because the nonlinearity of the electrostatic torque becomes more serious for larger total optical scan angles (or driving voltages). The potential application of this method includes LiDAR, which requires around 100 lines for the slow axis, although a 1% scan accuracy is not nearly precise enough for general display systems.

As a strict performance index, the USR was also estimated under a given criterion defined by the scan line error (SLE) as a percentile. For example, an SLE of 1% indicates that a certain scan line deviates from the intended trajectory by 1% of the total OSA. In other words, the USR is the longest scan range where the difference between the desired output and the scan output is less than 1% SLE. The estimated USR is shown in Figure 9 with respect to the total OSA at for SLEs of 1.0% and 1.5%.

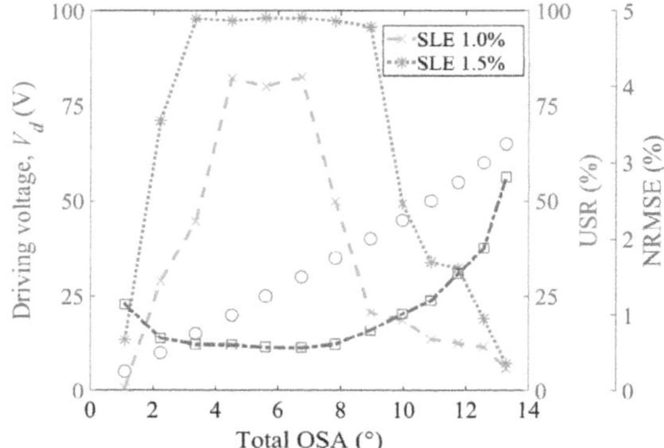

Figure 9. Experimental performance of the proposed input shaping with respect to the total optical scan angle (total OSA) (driving frequency: 20 Hz, sampling rate: 50 kHz): driving voltage with respect to total optical scan angle (circles in blue), usable scan range (dashed lines), and normalized root-mean-square error (NRMSE) (dash-dotted line in green).

When the SLE is 1.5%, the USR is 96% or higher for a value of total OSA ranging between 3.4° and 8.9°. For a small value of total OSA (1.2°–2.2°), the USR is poor, even though the shape of the scan output appears to be close to a triangular waveform. In cases with small scan angles, the high-frequency noise induced an additional degradation of 1.5% in SLE, as opposed to cases with large scan angles. This observation can be generally explained by the fact that the same amount of noise would degrade a USR with a small total OSA more than it would that with a large total OSA because the tolerable scan error becomes stricter for a smaller total OSA. Note that the NRMSE value shows a tendency to decrease with the USR. This implies that USR is valid as an effective performance index of the optical microscanner.

4.2. Driving Frequency

To examine the application limits of ISETF with respect to the driving frequency, the ISETF was experimentally implemented for driving frequencies from 20 Hz to 300 Hz. Generally, the higher the sampling rate, the more frequency components are included in the process of input shaping, resulting in the desired output. Thus, the sampling rate was fixed at 50 kHz to avoid the influence of the sampling rate on the performance. The driving voltage was also determined to generate a constant scan angle of 5.6° in the total OSA, at which one of the best USRs was achieved in Section 4.1.

Figure 10 compares the experimental scan output (solid line in black) and the ideal triangular output (dashed line in red) for various driving frequencies. When the frequency increases, the experimental scan output tends to deviate from the ideal triangular output. Particularly, the driving signals at the high frequencies tend to excite the natural oscillation mode at 14.6 kHz, which do not have a large enough amplitude to degrade the scan output.

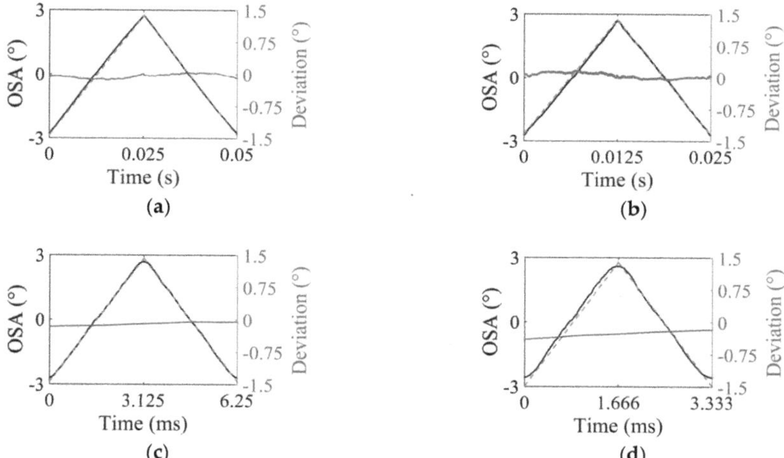

Figure 10. Performance of the ISETF with respect to the driving frequency (driving voltage: 25 V, sampling rate: 50 kHz): (**a**) 20 Hz, (**b**) 40 Hz, (**c**) 160 Hz, and (**d**) 300 Hz (scan output using the ISETF: solid line in black; ideal triangular output: dashed lines in red).

To quantitatively evaluate the scan outputs, the NRMSE and USR were analyzed with respect to the driving frequency, as shown in Figure 11. The total OSAs are also displayed, excluding the region with C-nonlinearity. The experimental total OSA varies slightly with respect to the driving frequency, despite the scanner being actuated by the driving voltage estimated to generate a constant total OSA value of 5.6°.

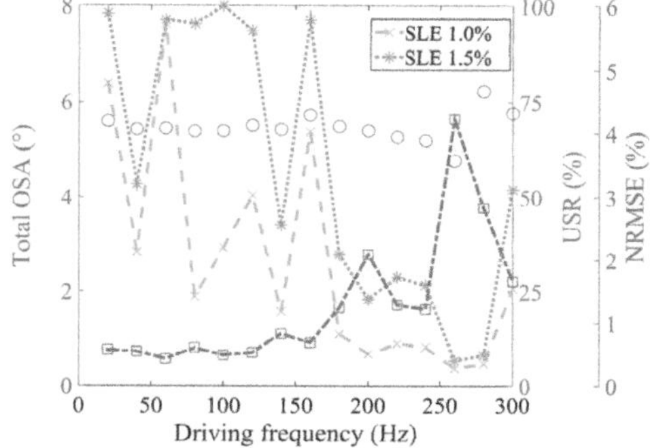

Figure 11. Experimental performance of the proposed input shaping with respect to driving frequency (driving voltage: 25 V, sampling rate: 50 kHz): total OSA with respect to driving frequency (circles in blue), usable scan range (dashed lines), and NRMSE (dash–dot line in green).

The NRMSE values remain below 1% for frequencies up to 160 Hz, which corresponds to the C-linear region for a scan angle of 5.6° or smaller. The NRMSE values abruptly degrade beyond the C-linear region.

The USR is 90% or higher for up to 160 Hz, except at two frequencies, when the SLE is 1.5%. To examine why the USR is specifically poor at the two frequencies of 40 Hz and 140 Hz, the frequency components of the triangular input were analyzed for the two driving frequencies and compared with the fundamental mode (264 Hz) of the microscanner. It was found that the fifth harmonic of 40 Hz and the first harmonic of 140 Hz are closer to the fundamental mode than those of other driving frequencies. These resonances are believed to be the major causes of the lower USR and larger NRMSE of the corresponding scan output.

There can be another possible explanation for the fact that the USRs are as poor as 55% or less, even if the shapes of the scan output are apparently close to the ideal triangular output. The poor USR might be partly attributed to the definition of USR that is determined by the region continuously satisfying the SLE requirement. In other words, there should be no single data point that deviates from the ideal triangular output beyond the SLE requirement. Indeed, although the USR is very practical as a performance index, it might sometimes lose its statistical meaning, unlike NRMSE.

For a driving frequency of 180 Hz or higher, the USRs fall off below 52% because the higher driving frequencies include a smaller number of harmonics in the input shaping calculation. The reduction in harmonics is attributed to the limitation from the Nyquist criterion, which results in the degradation of the triangular shape of the output. Note that the NRMSE value decreases with the USR, implying that the validity of the USR in the range of driving frequency is similar to that in the range of the OSA.

4.3. Comparison to Other Methods

As shown in Table 2, the performance of the ISETF is experimentally compared with those of other methods [9]; this paper contains the best results of input shaping for open-loop and closed-loop control. For a quantitative comparison of the proposed ISETF method to the conventional methods, the driving frequencies were normalized by the resonance frequencies of the scanners used in each experiment. Half of the angle error (peak-to-peak error) was also assumed to correspond to the NRMSE value.

Table 2. Performance comparison between open-loop control and the proposed ISETF.

Control method		Open-loop control [9]	This paper
Frequency of fundamental torsional mode (FTM)		120 Hz	264 Hz
Total optical scan angle (total OSA)		20°	5.6°
20–40 Hz	Peak-to-peak error	66.0 m°	31.9 m°
	Normalized peak-to-peak error	**0.33 %**	**0.57%**
60–220 Hz	Peak-to-peak error	500.0 m°	50.4 m°
	Normalized peak-to-peak error	**2.50%**	**0.90%**
Control method		**Closed-loop control [9]**	**This paper**
20–140 Hz	Peak-to-peak error	46.0 m°	31.4 m°
	Normalized peak-to-peak error	**0.23 %**	**0.56%**
160–220 Hz	Peak-to-peak error	790.0 m°	73.9 m°
	Normalized peak-to-peak error	**3.95 %**	**1.32 %**

Under these conditions, critical frequency can be defined based on relative superiority in performance. The normalized peak-to-peak error of the ISETF is slightly inferior to those of the open-loop and closed-loop control by 0.14% up to 40 Hz and 0.33% up to 155 Hz, respectively, in terms of average values; the normalized peak-to-peak error as a percentile are equivalent to 7.84 m° and 18.5 m° for the total OSA of 5.6°, respectively. For the driving frequencies from the critical frequencies up to 220 Hz, the normalized peak-to-peak error of the ISETF is smaller than those of the open-loop control and closed-loop control by 1.60% (89.6 m°) and 2.63% (147.3 m°), respectively, in terms of average values. The better performance of the ISETF is particularly revealed by the reduced normalized peak-to-peak error values: by 11.7% (655.2 m°) at 130 Hz and 6.1% (341.6 m°) at 220 Hz, respectively. This is attributed to the fact that the ISETF compensated both the fundamental mode and higher modes, whereas other control methods do not.

5. Conclusions

This method directly uses the experimental transfer function (ETF) with no dynamic modeling of the microscanner. The ISETF enables the ETF to include harmonics, higher modes, and/or sub-resonance so that the transfer function can be obtained more accurately than by the conventional open-loop control. The proposed ISETF method was confirmed to effectively remove residual oscillation caused both by the fundamental mode and by the higher modes of a microscanner. The limitations of the ISETF were experimentally examined with respect to the OSA and frequency, showing that a large USR was achievable for OSAs up to 8.9° and for driving frequencies in the range up to 160 Hz. From the experimental results, it is expected that the proposed ISETF can effectively reduce the residual oscillation caused by higher modes or crosstalk in two-axis driving compared to the conventional open-loop method. As a further study, electrodes with a larger thickness are being considered in experiments to extend the application of the ISETF to larger scan angles.

Author Contributions: K.K. conceptualized the methodology, performed the experiments, analyzed the results, and wrote the manuscript. S.M. designed and fabricated the microscanner. Y.P. contributed to the experiment. J.K. provided the advice in the calculation of the electrostatic torque. J.-H.L. supervised this research, analyzed the results, and reviewed the manuscript with substantial corrections and suggestions.

Funding: This work was supported in part by the Technology Innovation Program of the MOTIE (10065179), by the Unmanned Vehicles Advanced Core Technology Research and Development Program (NRF-2016M1B3A1A01937575), by the research project of WeMEMS Co (WP18-1) and by Technology Transfer and Commercialization Program through INNOPOLIS Foundation (2018-GJ-RD-0018-01-101), South Korea.

Conflicts of Interest: The authors declare no conflict of interest.

Abbreviations

ETF	Experimental transfer function
ISETF	Input shaping method based on an experimental transfer function
NRMSE	Normalized root-mean-square error
OSA	Optical scan angle
SLE	Scan line error
USR	Usable scan range

References

1. Jung, W.; McCormick, D.T.; Zhang, J.; Wang, L.; Tien, N.C.; Chen, Z. Three-dimensional endoscopic optical coherence tomography by use of a two-axis microelectromechanical scanning mirror. *Appl. Phys. Letters* **2006**, *88*. [CrossRef]
2. Hofmann, U.; Janes, J.; Quenzer, H.-J. High-Q MEMS resonators for laser beam scanning displays. *Micromachines* **2012**, *3*, 509–528. [CrossRef]
3. Milanović, V.; Kasturi, A.; Yang, J.; Hu, F. Closed-loop control of gimbal-less MEMS mirrors for increased bandwidth in LiDAR applications. In Proceedings of the Laser Radar Technology and Applications XXII 2017, Anaheim, CA, USA, 9–13 April 2017; p. 101910N.
4. Iyer, V.; Losavio, B.E.; Saggau, P. Compensation of spatial and temporal dispersion for acousto-optic multiphoton laser-scanning microscopy. *J. Biomed. Opt.* **2003**, *8*, 460–471. [CrossRef]
5. Lu, C.D.; Kraus, M.F.; Potsaid, B.; Liu, J.J.; Choi, W.; Jayaraman, V.; Cable, A.E.; Hornegger, J.; Duker, J.S.; Fujimoto, J.G. Handheld ultrahigh speed swept source optical coherence tomography instrument using a MEMS scanning mirror. *Biomed. Opt. Expr.* **2013**, *5*, 293–311. [CrossRef] [PubMed]
6. Schroedter, R.; Janschek, K.; Sandner, T. Jerk and current limited flatness-based open loop control of foveation scanning electrostatic micromirrors. *Proc. IFAC Vol.* **2014**, *47*, 2685–2690. [CrossRef]

7. Cogliati, A.; Canavesi, C.; Hayes, A.; Tankam, P.; Duma, V.F.; Santhanam, A.; Thompson, K.P.; Rolland, J.P. MEMS-based handheld scanning probe with pre-shaped input signals for distortion-free images in Gabor-domain optical coherence microscopy. *Opt. Expr.* **2016**, *24*, 13365–13374. [CrossRef]
8. Mohammadi, A.; Fowler, A.G.; Yong, Y.K.; Moheimani, S.O.R. A feedback controlled MEMS nanopositioner for on-chip high-speed AFM. *J. Microelectromech. Syst.* **2014**, *23*, 610–619. [CrossRef]
9. Schroedter, R.; Roth, M.; Janschek, K.; Sandner, T. Flatness-based open-loop and closed-loop control for electrostatic quasi-static microscanners using jerk-limited trajectory design. *Mechatronics* **2018**, *56*, 318–331. [CrossRef]
10. Borovic, B.; Liu, A.Q.; Popa, D.; Cai, H.; Lewis, F.L. Open-loop versus closed-loop control of MEMS devices: choices and issues. *J. Microelectromech. Microeng.* **2005**, *15*, 1917–1924. [CrossRef]
11. Pal, S.; Xie, H. Pre-shaped open loop drive of electrothermal micromirror by continuous and pulse width modulated waveforms. *IEEE J. Quantum Electron.* **2010**, *46*, 1254–1260. [CrossRef]
12. Krylov, S.; Maimon, R. Pull-in dynamics of an elastic beam actuated by continuously distributed electrostatic force. *J. Vib. Acoust.* **2004**, *126*, 332–342. [CrossRef]
13. Yu, S.; Piyabongkarn, D.; Sezen, A.; Nelson, B.J.; Rajamani, R.; Schoch, R.; Potasek, D.P. A novel dual-axis electrostatic microactuation system for micromanipulation. In Proceedings of the IEEE/RSJ International Conference on Intelligent Robots and System, Lausanne, Switzerland, 30 September–4 October 2002; pp. 1796–1801.
14. Horsley, D.A.; Wongkomet, N.; Horowitz, R.; Pisano, A.P. Precision positioning using a microfabricated electrostatic actuator. *IEEE T Magn.* **1999**, *35*, 993–999. [CrossRef]
15. Bazaei, A.; Maroufi, M.; Mohammadi, A.; Moheimani, S.O.R. Displacement sensing with silicon flexures in MEMS nanopositioners. *J. Microelectromech. Syst.* **2014**, *23*, 502–504. [CrossRef]
16. Moon, S.; Lee, J.; Yun, J.; Lim, J.; Gwak, M.-J.; Kim, K.-S.; Lee, J.-H. Two-axis electrostatic gimbaled mirror scanner with self-aligned tilted stationary combs. *IEEE Photon. Technol. Letters* **2016**, *28*, 557–560. [CrossRef]
17. Milanović, V. Linearized gimbal-less two-axis MEMS mirrors. In Proceedings of the Optical Fiber Communication Conference and National Fiber Optic Engineers Conference, San Diego, CA, USA, 22–26 March 2009; p. JThA19.

Article

Programmable Spectral Filter in C-Band Based on Digital Micromirror Device

Yunshu Gao [1,2], Xiao Chen [1,*], Genxiang Chen [1,*], Zhongwei Tan [2], Qiao Chen [1], Dezheng Dai [1], Qian Zhang [1] and Chao Yu [3]

1 College of Science, Minzu University of China, Beijing 100081, China; gaoyunshu@126.com (Y.G.); chenqiao0117@163.com (Q.C.); daidezheng@126.com (D.D.); qianzhang1521@163.com (Q.Z.)
2 School of Electronic and Information Engineering, Beijing Jiaotong University, Beijing 100044, China; zhwtan@bjtu.edu.cn
3 School of Electronic Engineering, Beijing University of Posts and Telecommunications, Beijing 100876, China; yu_chao@bupt.edu.cn
* Correspondence: xchen4399@126.com (X.C.); gxchen_bjtu@163.com (G.C.)

Received: 11 January 2019; Accepted: 26 February 2019; Published: 27 February 2019

Abstract: Optical filters have been adopted in many applications such as reconfigurable telecommunication switches, tunable lasers and spectral imaging. However, most of commercialized filters based on a micro-electrical-mechanical system (MEMS) only provide a minimum bandwidth of 25 GHz in telecom so far. In this work, the programmable filter based on a digital micromirror device (DMD) experimentally demonstrated a minimum bandwidth of 12.5 GHz in C-band that matched the grid width of the International Telecommunication Union (ITU) G.694.1 standard. It was capable of filtering multiple wavebands simultaneously and flexibly by remotely uploading binary holograms onto the DMD. The number of channels and the center wavelength could be adjusted independently, as well as the channel bandwidth and the output power. The center wavelength tuning resolution of this filter achieved 0.033 nm and the insertion loss was about 10 dB across the entire C-band. Since the DMD had a high power handling capability (25 KW/cm^2) of around 200 times that of the liquid crystal on silicon (LCoS) chip, the DMD-based filters are expected to be applied in high power situations.

Keywords: programmable spectral filter; digital micromirror device; optical switch

---

1. Introduction

To maximize the use of fiber spectral capacity and improve network efficiency, the International Telecommunication Union (ITU) G.694.1 standard has replaced the earlier G.692 specification to eliminate inefficient optical guard bands. The next generation optical network is a flexible elastic grid network that can dynamically allocate an amount of spectral resources as needed in 12.5 GHz increments for individual channels to support different symbol rates. It appears that traditional reconfigurable optical add/drop multiplexers (ROADMs) that provide optical switching of the fixed wavelength need to be upgraded to support flexible grid channels [1]. The flexible-grid ROADMs [2] must consist of programmable channel switching devices, such as wavelength selective switches (WSSs) or flexible-grid filters.

Currently there are two main competing technologies used in commercial WSSs and filters: The micro-electrical-mechanical system (MEMS) and the liquid crystal on silicon (LCoS) spatial light modulator (SLM). With the digital micromirror device (DMD) as an electrical input, optical output MEMS performs efficient and reliable spatial light modulation. The DMD developed by Texas Instruments is an array of thousands to millions of tiny highly reflective aluminum micromirrors which can be addressed independently. It has the advantage of low cost, high power handling and a fast frame rate, so that it has become an attractive solution for modulating spectrum resources in reconfigurable telecommunication switches [3,4], dynamic spatial and image filters [5,6], tunable lasers [7–10] and equalizers for erbium-doped fiber amplifiers (EDFAs) [11]. However, DMD and MEMS based reconfigurable optical filters [12–14] up to date have the minimum spectral bandwidth of 25 GHz in C-band (The TrueFlex Twin Multi-Cast Switch produced by Lumentum), so that they cannot adapt to the development of the new ITU-T standard grid frequencies. Although it was reported LCoS-based WSSs [15,16] have achieved 12.5 GHz or even smaller minimum bandwidth and higher tuning resolution, the LCoS processor under high power operation starts to deteriorate in function, even acquiring permanent, irreparable damage [17]. In comparison, the DMD has a high power handling capability and the only limit of the device under high power illumination is that the aluminum micromirrors must operate below 150 °C [18]. Therefore, it is expected to greatly broaden the application of DMD based optical filters in high power situations.

In this paper, we employ a DMD combining with a high-line-density transmission grating into a 2-f optical system to demonstrate the programmable spectral filter with flexible center wavelength, elastic bandwidth and high power handling. The minimum bandwidth achieved was 12.5 GHz. This optical filter can become a pre-filter to obtain target spectrum or an equalizer for EDFAs. It is important to compensate lower power handling of LCoS-based switches in an optical network.

2. System Design

Figure 1 illustrates the layout of the DMD-based optical filter which consists of a fiber-coupling microlens array, a polarization converter, two lenses, a transmission grating and a DMD. Two ports from the fiber-coupling microlens array with a 127 μm-pitch were used as an input and an output. For the high-line-density transmission grating with 1201.2 line/mm is S-polarization dependent, a polarization converter was inserted after the micro-lens to modulate the polarization state of an input beam. The DMD adopted in system consisted of 1024 × 768 mirrors on a pitch of δ = 13.68 μm with ±12° micromirror tilt by software control. It had a highly efficient steering of NIR light and an anti-reflection coated substrate which assured a front cover reflection of less than 0.5% between 1400 and 1700 nm.

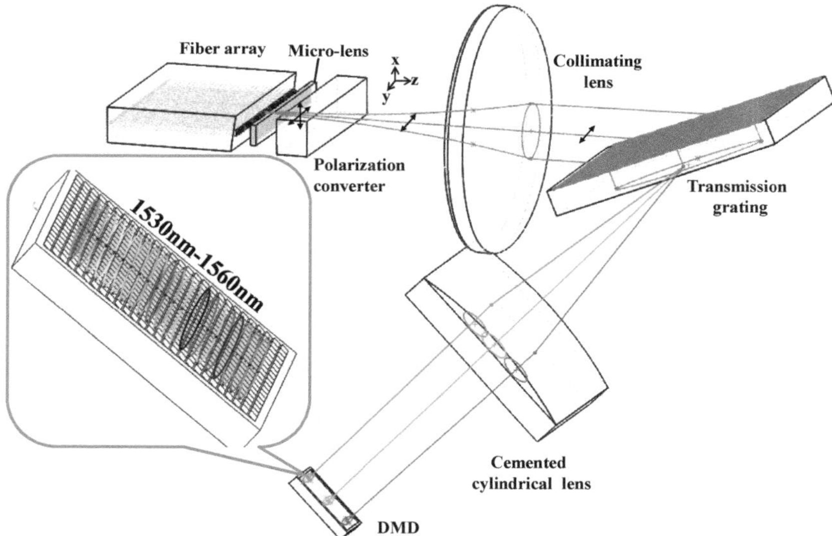

Figure 1. Diagram of programmable optical filter based on the digital micromirror device (DMD) chip.

Figure 2a is the xz-plane view and Figure 2b is the yz-plane view. We simplify the optical configuration by ignoring the angle of a transmission grating, so that the optical axis shown in Figure 2 is a straight line. A microlens combining with a collimating lens converted an input divergent Gaussian beam into a 6 mm-diam parallel beam [19,20]. The collimated broad-band beam was angularly dispersed in the x-axis direction by a grating and then focused into an elliptical spot on a different area of the DMD after a cylindrical lens. The purpose of the elliptical spot was to obtain the minimum bandwidth and high diffraction efficiency [16]. The DMD was placed at the focal plane of both the cylindrical lens (f_2 = 140 mm) and the collimating lens (f_1 = 300 mm) to realize the function that each micromirror was controlled at the on or off states to select and steer arbitrary wavebands precisely to the output.

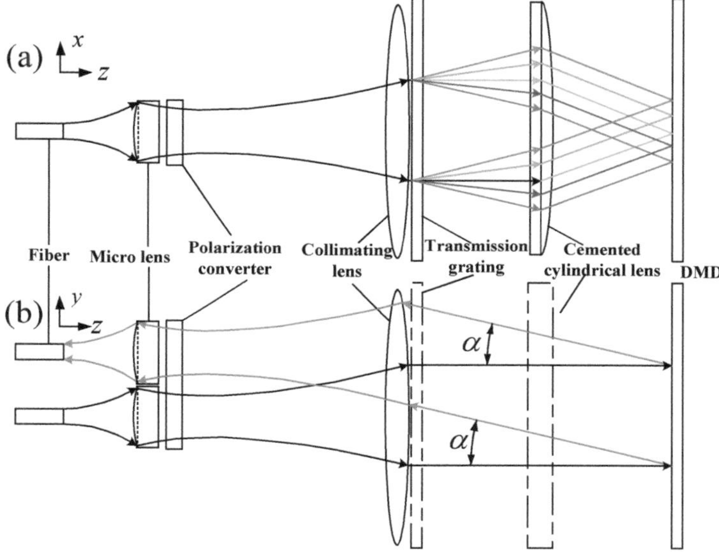

Figure 2. Layout of the filter optics: (**a**) The view in xz-plane showing light being de-multiplexed (**b**) The view in yz-plane showing light deflected by a DMD.

2.1. Diffraction Efficiency of DMD

In general, the long side of the DMD chip is aligned with the dispersion strip along the x-axis direction to maximize the spectrum utilization. The binary amplitude grating patterns are uploaded onto the DMD to control the corresponding micromirrors to tilt $\pm 12^\circ$ angle along their diagonals. The diffraction behavior of the several hundred thousand individually tilted micromirrors array as shown in Figure 3a is similar to a two-dimensional blazed grating. The diffraction distribution by the DMD and the coordinate system (x_0, y_0, z_0) is established in Figure 3b. The blue line represents an input beam and the red lines are the corresponding high-order diffraction beams in space. According to the 2D diffraction model [10], the diffraction angle of each order (p, q) is written as:

$$\begin{cases} \phi_{out}(p,q) = \tan^{-1}\left(\frac{H(q)}{G(p)}\right) \\ \theta_{out}(p,q) = \sin^{-1}\left(\frac{G(p)}{\cos\phi_{out}(p,q)}\right), \end{cases} \tag{1}$$

where $H(q) = \frac{q\lambda}{\delta} + \sin\theta_{in}\cos\phi_{in}$, $G(p) = \frac{p\lambda}{\delta} + \sin\theta_{in}\cos\phi_{in}$. ϕ_{in} is the incident angle between the input plane and y_0-axis, θ_{in} is the angle between the incident beam and z_0-axis in output plane as shown in Figure 3c. ϕ_{out} and θ_{out} are defined in the same way. In Equation (1), $(p, q) = 0, \pm 1, \pm 2, \cdots$

represent different diffraction orders. The diffraction distribution of an arbitrary order can be obtained when the incident angle ϕ_{in} and θ_{in} are provided.

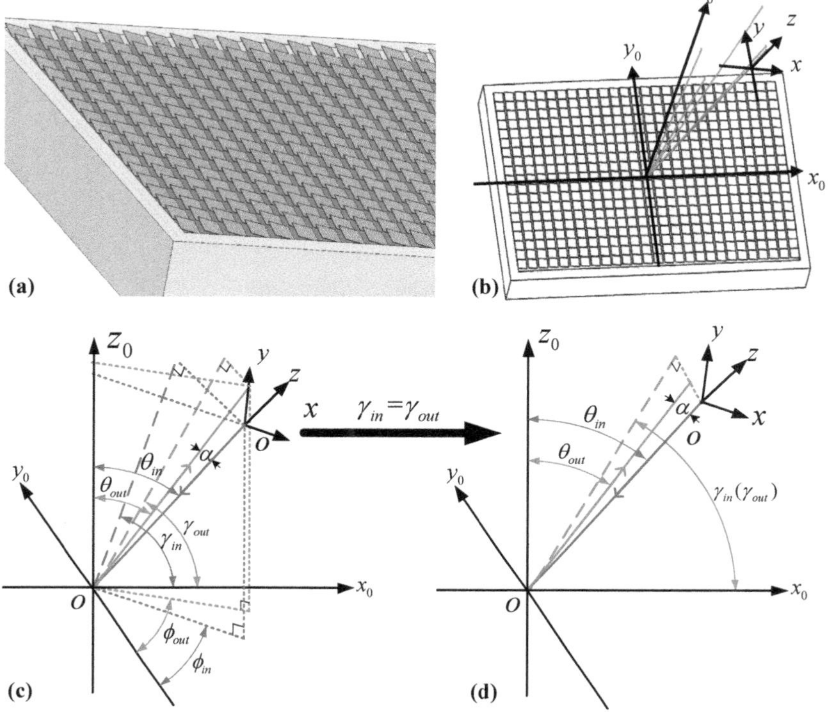

Figure 3. (**a**) Two-dimensional diffraction model of the DMD. (**b**) The incident beam and diffraction beam. (**c**) The coordinate system (x_0, y_0, z_0). (**d**) Distribution of the input and output beam when $\gamma_{in} = \gamma_{out}$.

As shown in Figure 3c, since the port distribution direction is perpendicular to the spectrum dispersion direction that is the x_0-axis direction, the incident angle needs to be adjusted to ensure the diffraction order with highest intensity is located in yoz-plane and routed back into an output port. The angle α is defined as that between input beam and diffraction beam. It is necessary to ensure $\gamma_{in} = \gamma_{out}$ in Figure 3d, with γ_{in} and γ_{out} being two angles between the projection of diffraction beam in x_0oz_0-plane and x_0-axis respectively. So the incident angles (θ_{in}, ϕ_{in}) satisfy the following condition:

$$\frac{\cos\theta_{in}}{\sin\theta_{in}\sin\phi_{in}} = \frac{\cos\theta_{out}}{\sin\theta_{out}\sin\phi_{out}}. \tag{2}$$

Based on the 2D diffraction model above, when a 1550 nm beam radiates on the +12° tilt DMD, it is noticed that the (−3,−3)-order diffraction beam always has a higher efficiency than the others. So multiple optimal incident angles according to Equations (1) and (2) are calculated and shown in Figure 4a. The angle α and normalized diffraction intensity of the (−3,−3)-order diffraction beam as a function of incident angles are presented in Figure 4b,c, respectively. Although the maximum normalized diffraction intensity can achieve 55%, the larger angle $\alpha = 2.5°$ worsens the optical aberration and insertion loss. So α is controlled at about 1° when θ_{in} = 13.91°, ϕ_{in} = 42°.

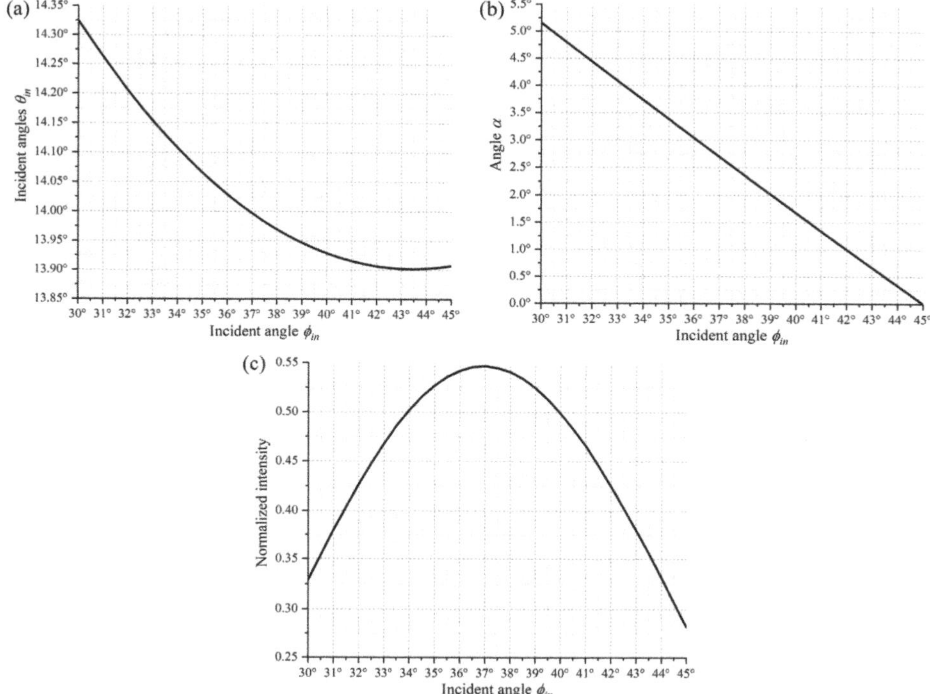

Figure 4. (**a**) Optimal incident angles(θ_{in}/ϕ_{in}) for the optical system. (**b**) Dependence of the angle α on the incident angles θ_{in}. (**c**) Dependence of the normalized intensity of output beam on incident angles θ_{in}.

Table 1 gives the diffraction principal maximum beam when a 1550 nm beam radiating on the DMD at θ_{in} = 13.91°, ϕ_{in} = 42°, including the corresponding diffraction angle $(\theta_{out}/\phi_{out})$ and relative intensity $I(\theta, \phi)$ of each (p, q)-order. The brightest order of diffracted light is $I(-3, -3)$ = 0.425, so that it is selected to couple into the output port while the other peaks are dramatically dropped out. The insertion loss by the DMD diffraction is around 3.7 dB.

Table 1. Irradiance maxima of light at 1550 nm radiating on the DMD over a large solid angle θ_{out}/ϕ_{out}, (I) of each diffraction order.

p/q	−4	−3	−2	−1
−4	22.8/45 (0.002)	18.6/30.4 (0.035)	16.1/9.9 (0.005)	-
−3	18.6/59.6 (0.019)	**13.2/45 (0.425)**	9.7/16.5 (0.05)	10.0/−22.1 (0.009)
−2	16.1/80.1 (0.011)	9.7/73.5 (0.213)	3.872/45 (0.018)	4.7/−54 (0.004)
−1	−16.4/−76.6 (0.001)	−10.0/−67.9 (0.03)	−4.7/−36.1 (0.003)	-

2.2. Power Handling of Optical filter

The power handling is one of the important specifications of optical filters. When the continuous wave (CW) laser illuminates a DMD, excessive energy absorption by on-surface aluminium-mirrors generally leads to the abnormal operation or even irreversible damage of the device. Therefore, it is necessary to keep the operating temperature below a critical point of 150 °C, and the average intensity cannot exceed 25 KW/cm^2 in the visible band [18]. In general, the damage threshold depends on the illumination wavelength and intensity profile. For example, a damage threshold for 1550 nm is twice of 645 nm. As shown in Figure 5, a Gaussian beam has a maximum power density twice of the uniform beam when both beams have the same spot size and power. It is reported the damage threshold of the DMD for 1550 nm Gaussian beam is estimated to be about 25 KW/cm^2 for CW-laser. Faustov [21] demonstrated that the measured threshold is up to 22 mW corresponding to 12 KW/cm^2 when a He-Ne laser at 633 nm is focused onto a 13.7 $\mu m \times 13.7$ μm-size micromirror. Furthermore, when an input laser at 1064 nm is below 30 mW (21 KW/cm^2), micromirrors do not exhibit any visible damage. Schwarz et al. [22] also experimentally showed the damage threshold of a 19.3 KW/cm^2 by 532 nm CW-laser.

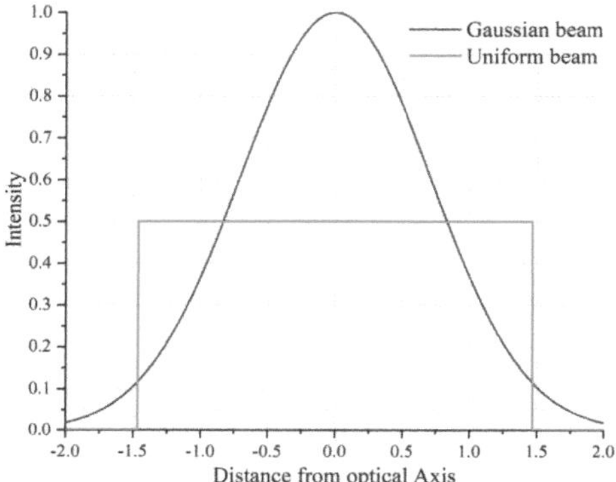

Figure 5. The beam intensity distribution of the Gaussian beam and uniform beam.

The compressed light spot on the DMD means not only a narrower bandwidth for the filter, but also a higher energy density. The spot size on the DMD for this filter is measured to be 60 $\mu m \times 9$ mm, so the max input power is about 135 W (50 dBm) corresponding to 25 KW/cm^2. The power handling of commercialized WSSs (Waveshaper 16,000 A produced by Finisar Corporation) and LCoS based filters is 27 dBm input power in maximum. Therefore, DMD-based filters are an irreplaceable solution in high power situation.

3. Experimental Results and Discussion

Figure 6 is the arrangement of the optical filter in experiment. The amplified spontaneous emission (ASE) light source in 1530–1560 nm was injected into the system as input signals. An optical spectrum measurement analyzer AQ6370C-YOKOGAWA was applied to measure the insertion loss, 3 dB-bandwidth and tuning resolution of the central wavelength. Figure 7 shows that the total loss was around 10 dB across the entire C-band with the ripples of 0.5 dB caused by the gap between micromirrors. The insertion loss mainly included 1.5 dB from the fiber-coupling microlens array, 0.5 dB from the polarization converter, 1.0 dB from the transmission grating and 4 dB from the DMD. In addition, when incident angle of the input beam at θ_{in} = 13.91°, ϕ_{in} = 42°, a mismatch between the focal plane of the cylindrical lens and the DMD caused 3 dB extra insertion loss. Although this 3 dB loss caused by oblique incident beam could be avoided by replacing the DMD with a 10.8 μm-pitch micromirror that had a 98% diffraction efficiency to vertical incident beam [10], it was not applied in the NIR-band without an anti-reflection coating, which would introduce more loss. The measured intrinsic polarization dependent loss (PDL) within the 12.5 GHz-bandwidth was less than 1 dB for the optical system. As the signal beam was diffracted by the transmission grating, a combination of conical diffraction and optical aberrations lead to the fluctuation of insertion loss of about 1 dB [16].

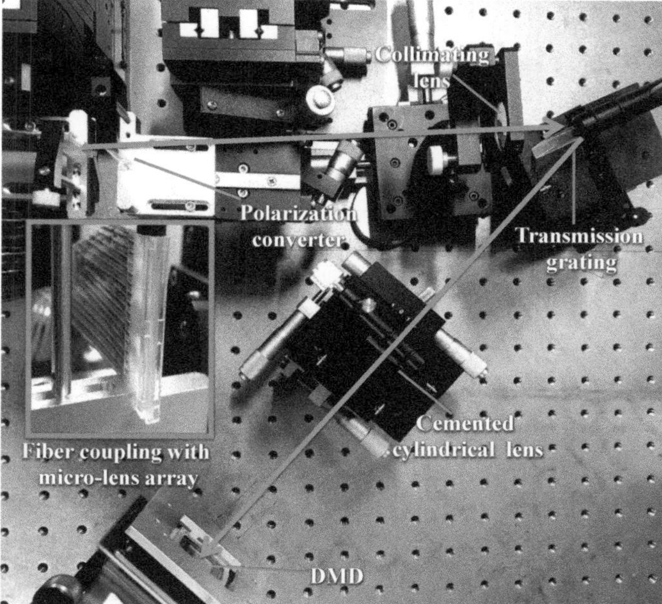

Figure 6. Arrangement of tunable optical filter.

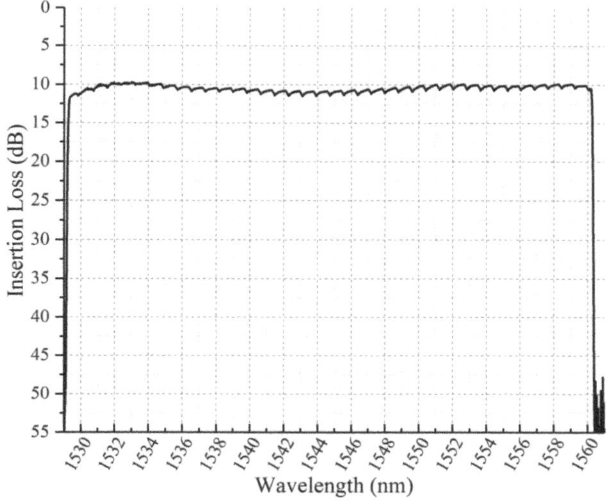

Figure 7. Total insertion loss as a function of C-band wavelength in filter system.

Figure 8a shows that the center wavelength could be tuned flexibly in step of 0.033 nm with the 3 dB-bandwidth of 12.5 GHz. In Figure 8b, arbitrary wavebands could be filtered and configured at a minimum resolution of 0.033 nm. The 3 dB-passband could be adjusted flexibly from 12.5 GHz to 50 GHz by a step of 12.5 GHz in Figure 8c. The minimum filter bandwidth could achieve 12.5 GHz, however it was noticed that the top of wavelength profile was not flat enough. The measured passband at nine ITU-T G.694.1 standard grid frequencies with 25 GHz channel separation had about -15 ± 1 dB channel crosstalk in Table 2 and Figure 8d. Since the power falling edge of this filter was about 12 GHz spectral width in 20 dB, very narrowing spectral guard bands with small crosstalk could be set for 50 GHz and 100 GHz spaced ITU-T channels in the C-band.

Table 2. Channel crosstalk and offset level at nine grid frequencies (25 GHz channel separation).

Center Wavelength (nm)	1550.09	1550.30	1550.50	1550.71	1550.91	1551.11	1551.32	1551.53	1551.74
Offset Level (dB)	−0.557	−0.25	−0.067	−0.029	−0.502	−0.267	−0.124	0	−0.419
Channel Crosstalk (dB)	−16.345	−16.46	−16.132	−14.677	−14.223	−15.93	−15.739	−15.677	−16.319

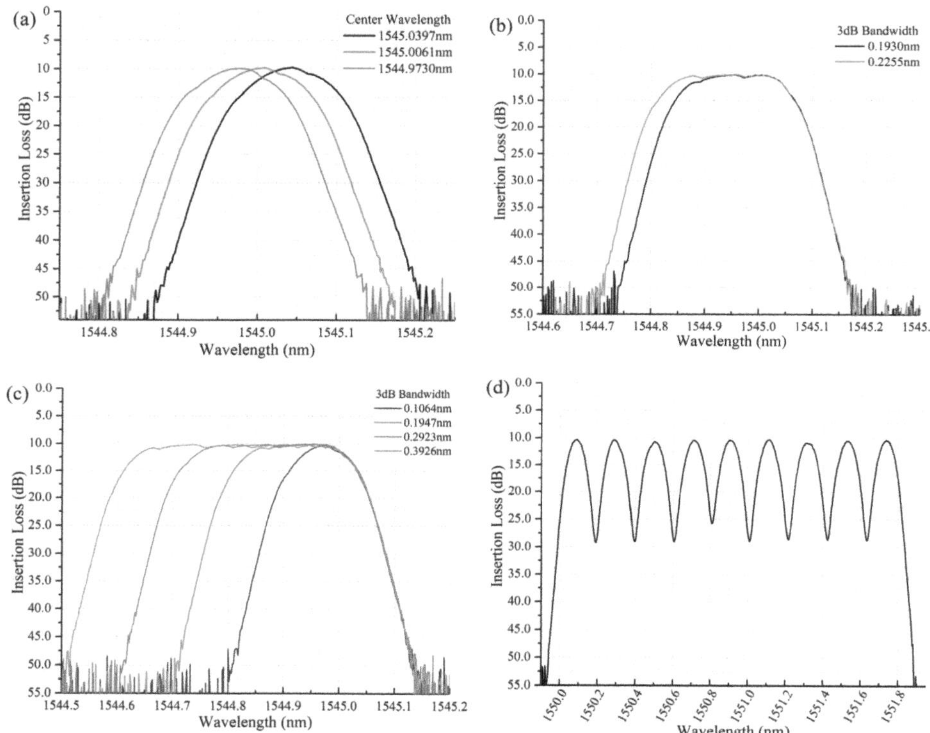

Figure 8. (**a**) The minimum tuning resolution of center wavelength in the filter. (**b**) The minimum tuning step of spectral bandwidth. (**c**) 3 dB-bandwidth from 12.5 GHz to 50 GHz with a step of 12.5 GHz. (**d**) Measured passband at 9 G.694.1 standard grid frequencies with 25 GHz channel separation.

In Figure 9a, the optical filter also provided a function of optical power attenuation. It was realized by controlling the corresponding micromirrors number in different locations to modulate the output luminous flux. The optical power attenuation could be adjusted from 0 dB to 40 dB flexibly with a resolution of 0.5 dB. Figure 9b shows the micromirrors information used to control the optical attenuation. This filter had 50 dBm maximum input power, and was an excellent equalizer for high power erbium-doped fiber amplifiers.

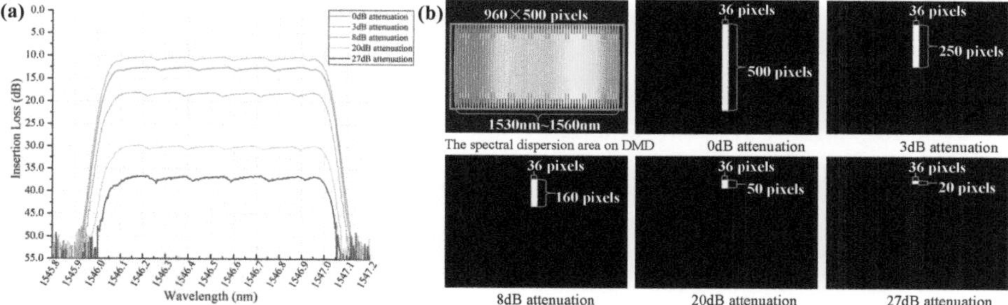

Figure 9. Optical power attenuation of the optical filter and the corresponding binary image.

4. Conclusions

We propose and demonstrate a tunable optical filter with max 50 dBm input power, flexible central wavelength and bandwidth by employing a DMD processor into the system. The total insertion loss of this filter was about 10 dB across the entire C-band. The center wavelength and bandwidth of multi-channel could be tuned in the step of 0.033 nm independently. Although the minimum bandwidth could achieve 12.5 GHz, the performance of the channel crosstalk for 12.5 GHz and 25 GHz ITU grid especially still needs further improvement, as the spectral does not have an ideal flat-topped profile. In future work, we plan to optimize the minimum bandwidth by employing a specially designed cylindrical lens system to eliminate the spherical aberration and chromatic aberration, and further decrease the spot size in the x-axis on the DMD.

Author Contributions: Conceptualization, G.C.; formal analysis, Y.G., Q.C., Z.T. and D.D.; investigation, C.Y. and Q.Z.; methodology, X.C.; project administration, X.C.; writing—original draft preparation, Y.G.; writing—review and editing, X.C.

Acknowledgments: This work was supported by the National Natural Science Foundation of China Grant number 61627814, 61275052.

Conflicts of Interest: The authors declare no conflict of interest. The funders had no role in the design of the study; in the collection, analyses, or interpretation of data; in the writing of the manuscript, or in the decision to publish the results.

References

1. Anuj Malik, G.H. The Evolution of Next-Gen Optical Networks: Terabit Super-Channels and Flexible Grid ROADM Architectures. In Proceedings of the SCTE Cable-Tec Expo 2014, Denver, CO, USA, 22–25 September 2014.
2. Woodward, S.L.; Feuer, M.D. Benefits and requirements of flexible-grid ROADMs and networks. *IEEE/OSA J. Opt. Commun. Netw.* **2013**, *5*, A19–A27. [CrossRef]
3. Blanche, P.A.; Carothers, D.; Wissinger, J.; Peyghambarian, N. DMD as a diffractive reconfigurable optical switch for telecommunication. In Proceedings of the Emerging Digital Micromirror Device Based Systems and Applications V, San Francisco, CA, USA, 5–6 February 2013.
4. Knapczyk, M.T.; de Peralta, L.G.; Bernussi, A.A.; Temkin, H. Reconfigurable add–drop optical filter based on arrays of digital micromirrors. *J. Lightw. Technol.* **2008**, *26*, 237–242. [CrossRef]
5. Jin, D.; Zhou, R.; Yaqoob, Z.; So, P.T. Dynamic spatial filtering using a digital micromirror device for high-speed optical diffraction tomography. *Opt. Express* **2018**, *26*, 428–437. [CrossRef] [PubMed]
6. Hoover, R.; Henry, A.; Arce, G.R. DMD-based implementation of patterned optical filter arrays for compressive spectral imaging. *J. Opt. Soc. Am. A Opt. Image Sci. Vis.* **2015**, *32*, 80–89.
7. Liu, W.; Fan, J.; Xie, C.; Song, Y.; Gu, C.; Chai, L.; Wang, C.; Hu, M. Programmable controlled mode-locked fiber laser using a digital micromirror device. *Opt. Lett.* **2017**, *42*, 1923–1926. [CrossRef] [PubMed]
8. Luo, D.; Taphanel, M.; Längle, T.; Beyerer, J. Programmable light source based on an echellogram of a supercontinuum laser. *Appl. Opt.* **2017**, *56*, 2359–2367. [CrossRef] [PubMed]

9. Wood, T.C.; Elson, D.S. A tunable supercontinuum laser using a digital micromirror device. *Meas. Sci. Technol.* **2012**, *23*, 105204. [CrossRef]
10. Chen, X.; Yan, B.B.; Song, F.J.; Wang, Y.Q.; Xiao, F.; Alameh, K. Diffraction of digital micromirror device gratings and its effect on properties of tunable fiber lasers. *Appl. Opt.* **2012**, *51*, 7214–7220. [CrossRef] [PubMed]
11. Gu, C.; Chang, Y.; Zhang, D.; Cheng, J.; Chen, S.C. Femtosecond laser pulse shaping at megahertz rate via a digital micromirror device. *Opt. Lett.* **2015**, *40*, 4018–4021. [CrossRef] [PubMed]
12. Khan, S.A.; Riza, N.A. Demonstration of the MEMS Digital Micromirror Device-Based Broadband Reconfigurable Optical Add–Drop Filter for Dense Wavelength-Division-Multiplexing Systems. *J. Lightw. Technol.* **2007**, *25*, 520–526. [CrossRef]
13. Knapczyk, M.; Krishnan, A.; de Peralta, L.G.; Bernussi, A.; Temkin, H. High-resolution pulse shaper based on arrays of digital micromirrors. *IEEE Photonics Technol. Lett.* **2005**, *17*, 2200–2202. [CrossRef]
14. Knapczyk, M.; Krishnan, A.; de Peralta, L.G.; Bernussi, A.; Temkin, H. Reconfigurable optical filter based on digital mirror arrays. *IEEE Photonics Technol. Lett.* **2005**, *17*, 1743–1745. [CrossRef]
15. Xie, D.; Wang, D.; Zhang, M.; Liu, Z.; You, Q.; Yang, Q.; Yu, S. LCoS-based wavelength-selective switch for future finer-grid elastic optical networks capable of all-optical wavelength conversion. *IEEE Photon. J.* **2017**, *9*, 1–12. [CrossRef]
16. Gao, Y.; Chen, G.; Chen, X.; Zhang, Q.; Chen, Q.; Zhang, C.; Tian, K.; Tan, Z.; Yu, C. High-Resolution Tunable Filter With Flexible Bandwidth and Power Attenuation Based on an LCoS Processor. *IEEE Photonics J.* **2018**, *10*, 1–8. [CrossRef]
17. Carbajo, S.; Bauchert, K. Power handling for LCoS spatial light modulators. In Proceedings of the Laser Resonators, Microresonators, and Beam Control XX, San Francisco, CA, USA, 29 January–1 February 2018.
18. Laser Power Handling for DMDs. Available online: http://www.ti.com/lit/wp/dlpa027/dlpa027.pdf (accessed on 26 February 2019).
19. Iwama, M.; Takahashi, M.; Kimura, M.; Uchida, Y.; Hasesawa, J.; Kawahara, R.; Kagi, N. LCOS-based flexible grid 1 × 40 wavelength selective switch using planar lightwave circuit as spot size converter. In Proceedings of the 2015 Optical Fiber Communications Conference and Exhibition (OFC), Los Angeles, CA, USA, 22–26 March 2015; pp. 1–3.
20. Suzuki, K.; Ikuma, Y.; Hashimoto, E.; Yamaguchi, K.; Itoh, M.; Takahashi, T. Ultra-high port count wavelength selective switch employing waveguide-based I/O frontend. In Proceedings of the Optical Fiber Communication Conference, Los Angeles, CA, USA, 22–26 March 2015.
21. Faustov, A.R.; Webb, M.R.; Walt, D.R. Note: Toward multiple addressable optical trapping. *Rev. Sci. Instrum.* **2010**, *81*, 026109. [CrossRef] [PubMed]
22. Schwarz, B.; Ritt, G.; Koerber, M.; Eberle, B. Laser-induced damage threshold of camera sensors and micro-optoelectromechanical systems. *Opt. Eng.* **2017**, *56*, 034108. [CrossRef]

 *micromachines* 

Article

The Improvement on the Performance of DMD Hadamard Transform Near-Infrared Spectrometer by Double Filter Strategy and a New Hadamard Mask

Zifeng Lu [1,2], Jinghang Zhang [1], Hua Liu [1,2,*], Jialin Xu [3] and Jinhuan Li [1,2]

1 Center for Advanced Optoelectronic Functional Materials Research, and Key Laboratory for UV-Emitting Materials and Technology of Ministry of Education, Northeast Normal University, 5268 Renmin Street, Changchun 130024, China; luzf934@nenu.edu.cn (Z.L.); jinghang1927@163.com (J.Z.); lijh248@nenu.edu.cn (J.L.)
2 Demonstration Center for Experimental Physics Education, Northeast Normal University, 5268 Renmin Street, Changchun 130024, China
3 Changchun Institute of Optics, Fine Mechanics and Physics, Chinese Academy of Sciences, Changchun 130033, China; xujialinseu@163.com
* Correspondence: liuhua_rain@aliyun.com; Tel.: +86-180-0443-0180

Received: 7 December 2018; Accepted: 15 February 2019; Published: 23 February 2019

Abstract: In the Hadamard transform (HT) near-infrared (NIR) spectrometer, there are defects that can create a nonuniform distribution of spectral energy, significantly influencing the absorbance of the whole spectrum, generating stray light, and making the signal-to-noise ratio (SNR) of the spectrum inconsistent. To address this issue and improve the performance of the digital micromirror device (DMD) Hadamard transform near-infrared spectrometer, a split waveband scan mode is proposed to mitigate the impact of the stray light, and a new Hadamard mask of variable-width stripes is put forward to improve the SNR of the spectrometer. The results of the simulations and experiments indicate that by the new scan mode and Hadamard mask, the influence of stray light is restrained and reduced. In addition, the SNR of the spectrometer also is increased.

Keywords: spectrometer; infrared; digital micromirror device (DMD); signal-to-noise ratio (SNR); stray light

---

1. Introduction

In the 1970s, the Hadamard transform (HT) was proposed and developed into a relatively mature theory [1]. With the emergence of the mechanical encoding mask, the HT was applied to the near-infrared (NIR) spectrometer. The encoding mask is a key device in spectrometers. However, adopting the mechanical mask, the spectrometer exhibits a complex structure, low resolution, and short life. Compared with the traditional instrument, it possesses no advantage. The development of HT spectrometers is restricted by the encoding mask. Later, the digital micromirror device (DMD) was developed and applied to the HT spectrometer as an encoding mask. Because the HT spectrometer based on the DMD has several advantages such as a higher signal-to-noise ratio (SNR), wider spectral range, and low cost [2,3], DMD-based HT spectrometers have attracted significant research attention.

At present, the performance of HT spectrometers has been greatly improved, but they still have defects, such as the grating diffraction of the spectrometer, the two-dimensional grating diffraction of the DMD, and the poor spectral efficiency of the light source; these defects can make the spectral energy distribution uneven. Thus, the influence of stray light on the absorbance of the whole spectrum is varied; the lower the spectral energy is, the greater the influence by stray light is. The low-energy spectral band exhibits a low SNR, nonlinearity, whereas the high-energy spectral band performs well in those aspects. To improve the energy of the entire spectrum, Wang and colleagues proposed a spectrum-folded structure of a HT spectrometer and a special illumination optical device, but the structure of the spectrometer was complex [4,5]. Zhang et al. proposed a new algorithm to realize energy compensation of the spectrum and analyzed the effect of the HT on the noise without considering the noise distribution [6]. Quan et al. analyzed the spectral distortion in the HT spectrometer and presented a correction approach [7]. However, for the analysis of stray light, especially that with a high correlation of energy distribution, their processing effect was unsatisfactory. Xu et al. analyzed the influence of the HT on the noise before and after coding. They also proposed a new encoding mask to correct the anomaly in the spectra caused by optical defects [8]. With the variation in the height of the stripes, their new encoding mask exhibited a low utilization rate of the DMD.

To improve the performance of the DMD HT NIR spectrometer, a new method of the split waveband scanning is proposed in this paper to mitigate the impact of the stray light. It can not only reduce the influence of stray light on the low-energy spectral bands, but also improve the linearity and accuracy of absorbance in the low-energy spectral bands. On the other hand, a new Hadamard mask of variable width-stripe matching with each scanning area is presented to improve SNR. Based on the new scanning method and coding mask, the simulation and experimental results indicate that the stray light is suppressed and the spectral energy distribution is more uniform. The SNR of the spectrum is also improved, especially in the low-energy spectral band the SNR is increased significantly. It is demonstrated that by the proposed approach, the minimum SNR in the low-energy spectral band is improved by a factor of 7.434 greater than that of the traditional HT method.

2. Theory of Hadamard Transform (HT) Spectrometer with Digital Micromirror Device (DMD)

A schematic of the spectrometer designed by us is illustrated in Figure 1a. The incident light emitted from the sample pool is dispersed by the grating, and the dispersion spectrum imaged on the DMD plane by the imaging lens is encoded and reflected. Then, the reflected light is focused onto the detector by the converging lens. Finally, the detector signal is decoded and processed by a computer. Because DMD is programmable, multiple scan modes are available to the spectrometer such as the column scan mode, the Hadamard scan mode, and other multiplexed scan mode. To our spectrometer, the major scan mode is Hadamard scan. The coding matrix of the Hadamard spectrometer is an S-matrix determined by quadratic residue method and can be used to describe the patterns to be displayed on the DMD [9]. By this approach, the spectrum can be modified.

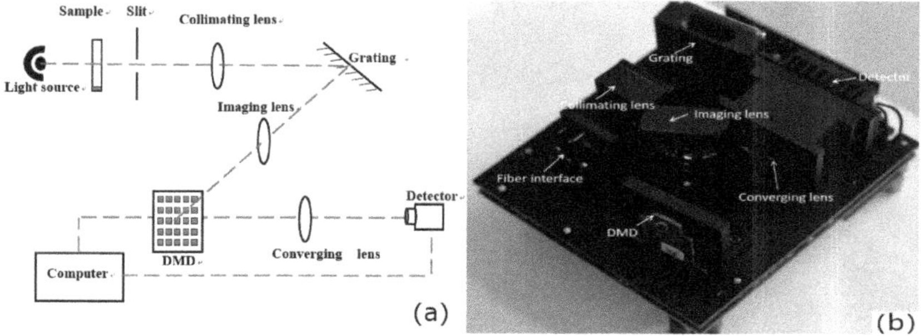

Figure 1. (**a**) Optical system of Hadamard transform (HT) spectrometer. (**b**) Optical structure of HT spectrometer.

Based on our theory, the HT NIR spectrometer with a DMD can be realized. The photograph of the spectrometer is shown in Figure 1b. The parameters of the spectrometer are listed in Table 1. Before the spectrometer is used, the calibration must be performed to get the relationship between the wavelength and the pixel location in the DMD column. Three lasers whose wavelengths are respectively 1550 nm, 1625 nm, and 2210 nm are used to calibrate approximately the spectrometer, and a Xenon Calibration Light Source of Ocean Optics is used to perform the accuracy calibration. The detailed calibration process of the spectrometer can be found in reference [10].

Table 1. Parameters of the HT near-infrared (NIR) spectrometer.

Components	Model/Parameters	Technical Indexes	Parameters
Light source	Tungsten lamp	Spectral range	1350–2450 nm
Slit	50 μm	Spectral resolution	$\leq$ 8 nm
Grating	200 lines/mm	DMD resolution	912 × 1140
DMD	DLP4500NIR	Sampling rate	1 s
DMD mirrors	Aluminum micromirror	-	-
Optical fibers	Infrared quartz	-	-
Detector	InGaAs $R = 3$ mm	-	-

The spectrum of the light source tested by the spectrometer designed by us is presented in Figure 2. The relative spectral energy is high in the range of 1350–2200 nm and low in the range of 2200–2450 nm. The spectral curves tested by some other spectrometers with different designs may have some differences, but the spectral distribution is the same. The energy in the central wavelength band of the spectrum is high and those on the two edges are low.

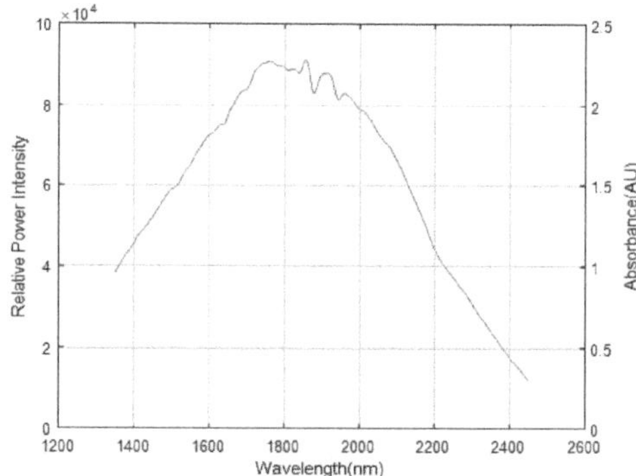

Figure 2. Light source spectrum.

3. The Measurement Result Increase in the Low-Energy Spectral-Band Absorbance

3.1. *Impact of Stray Light on the Spectral Band with Different Energies*

Stray light is an important measurement parameter of the spectrometers. The existence of the stray light results in the significant measurement errors [11]. Especially to the dispersive spectrometer, it may cause the nonlinear problems of the instrument. Thus, it is critical to suppress the stray light. In this study, the stray light originates mainly from the scattered light and reflected light of the spectrometer. When the detector receives a certain wavelength signal, it always mixes with some stray light, which is not part of the signal. The existence of stray light will reduce the measured absorbance, especially in spectral bands with strong absorption [8].

This type of stray light is very complex. It may lead to a test result with a homogeneous background. The background values can be measured before beginning the spectral test. When the DMD is closed, the signals received by the detector are the background values. By subtracting the background values from the spectral test signals, the influence of a large proportion of stray light can be corrected. Thus, its performance optimization may be less expensive than other spectrum analysis systems. Beyond that, there is still a fraction of stray light remaining in the energy distribution of the spectrum that cannot be corrected by this method. Using the Zemax software, we can build a nonsequence mode of the optical structure of the spectrometer. By this mode, the energy distribution of the slit images can be obtained in different wavelength bands on the surface of the DMD. The irradiance of the slit image at a wavelength of 1600 nm is shown in Figure 3. In this Figure, the energy of the slit image is the highest and there exist other stray lights with a lower energy around the slit image, except for the background light. Because of the secondary reflection of the devices in the spectrometer, there are two energy circles around the slit image, which may be related to the energy distribution of the slit. The secondary reflection mainly comes from the front and rear surfaces of the DMD window. In addition, ghosting can be observed under the slit image, which is formed by the window of the DMD. Because the micromirrors of DMD are easily broken, the optical window is a must to protect them from the surroundings and permit the light of a certain wavelength range to transmit. The material of the DMD window is a kind of optical glass (Corning 7056) whose main constituent is SiO_2. For visible, near infrared and ultraviolet wavebands, the anti-reflection films covered on the glasses are different. The refractive index of this type of glass is 1.487 at the wavelength of 545 nm, and the cutoff wavelength for transmission is 2.7 μm. Because the transmission of the DMD window in 1700–2500 nm is reduced, the absorption and reflection are strengthened. So, ghosting can't be avoided in a carefully designed spectrometer because of the characteristics of DMD. The stray lights formed by the secondary reflection have a great influence on the absorbance measurements of the low-energy spectrum [7].

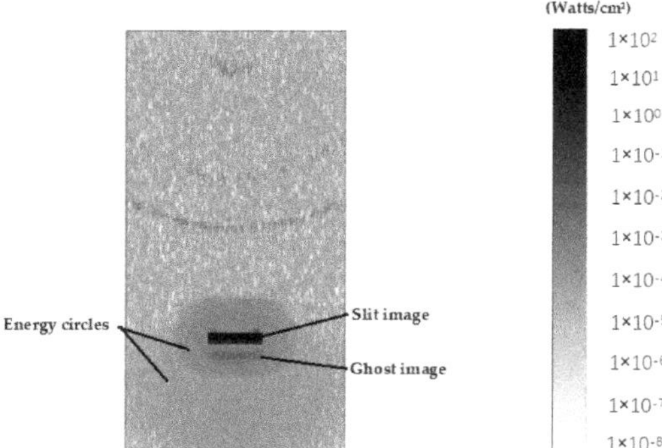

Figure 3. Incoherent radiation of slit image.

Therefore, we conduct some analyses to determine the influence of stray light. The absorption spectrum of 95% ethanol is respectively measured by our spectrometer and UV-Visible/NIR Spectrophotometer UH4150 produced by Hitachi High Technologies Corporation (Tokyo, Japan), as shown in Figure 4. From Figure 4, we can see some stronger absorption peaks of the C–H, $C–H_2$, and O–H bending and harmonic vibration in ethanol always exist in the long-wavelength band. However, some peak values are less accurate than the reference and some weak peaks cannot be observed. In 2200–2450 nm, the three peaks shown by black line B move about 1–1.5 nm in the direction of long wave and two peaks at 2280 and 2297 nm are not observed compared with the blue line A; In 1400–2200 nm, there is also a red shift of 1.3–3 nm at all peaks except at 1693, 1760, and 1937 nm compared with the reference spectrum. It may be caused by the inaccurate calibration of the spectrometer and the lack of the data points collected by our spectrometer. Comparing the measured spectrum with the reference spectrum, it shows some deviations of absorption intensity between the test values and the reference in whole wavelength band, and the difference of absorption intensities in 2200–2450 nm is greater than that in 1400–2200 nm. An important factor is the stray light. We suppose that when testing the sample spectrum, the low-energy waveband might be disturbed by the stray light from the high-energy waveband, and the high energy waveband will be affected by the stray light from the low energy waveband. The influence of the test results cannot be ignored.

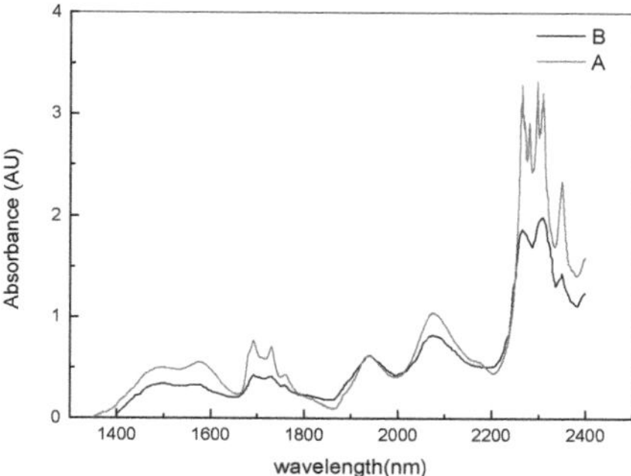

Figure 4. Absorbance spectrum of 95% ethanol. The blue line A is the reference spectrum tested by the UV-visible/NIR spectrophotometer UH4150. The black line B is the absorbance curve of the tested by the traditional scan method.

Then, we perform simulation analysis to verify this hypothesis. The light source is absorbed by the sample. The transmission spectrum is received by the detector, and the energy of the spectrum contains two parts. One is from the transmission light (T_λ) and the other is stray light (S_λ). The absorbance of a certain wavelength of the spectrum can be calculated by the following equations:

$$\mathrm{A} = \lg\left(\frac{I_\lambda}{T_\lambda + S_\lambda}\right), \quad (1)$$

$$S_\lambda = \sum_m k_m T_{\lambda m}, \quad (2)$$

where I_λ is the spectral intensity at wavelength λ and S_λ denotes the total stray light intensity of the high-energy wavelength band (or low-energy wavelength band). $T_{\lambda m}$ is the corresponding intensity of the sampling points of the transmission light and m is the spectral sampling number. k_m is the proportion coefficient between the stray light intensity and spectral intensity at a given wavelength. Because each proportion coefficient k_m is different and complex, we perform simulation experiments under the ideal condition, which state that in the same spectral band, the proportion coefficient k_m of the stray light intensity of each sampling point is replaced by the same coefficient k (k = 0.01, 0.001, 0.0001). To simplify the analysis process, the spectrum is divided into two parts: the high-energy wavelength band (1350–2200 nm) and low-energy wavelength band (2200–2450 nm). By this analysis, we suppose that the stray lights produced by the high-energy wavelength band and low-energy wavelength band have different impacts on each other.

First, we analyze the influence of stray light on the low-energy wavelength band. According to the spectrum presented in Figure 2, the range of spectral intensity (I_λ) is between 10,000 and 40,000. Assume that the absorbance value of low-energy wavelength band is fixed, and set it to be 1. Then the intensity of transmission light at a given wavelength: $T_\lambda = 0.1 I_\lambda$. In the high-energy wavelength band, the range of T_λ is between 40,000 and 80,000. By solving Equations (1) and (2), the curve of the intensity effect on the absorbance obtained by the simulation is shown in Figure 5. The purple line is the real absorbance. We can see that the lower the transmission light intensity is, the greater the influence of stray light from the high-energy wavelength band is. The tested values deviate from the real value. Next, we analyse the influence of stray light on the high-energy wavelength band. The range of spectral intensities (I_λ) is between 40,000 and 80,000. Assume that the absorbance value of high-energy wavelength band is fixed, and set it to be 0.155. Then the intensity of transmission light at a given wavelength: $T_\lambda = 0.7 I_\lambda$. In the low-energy wavelength band, the range of T_λ is between

10,000 and 40,000. The curve of the intensity that affects the absorbance obtained by the simulation is shown in Figure 6. From the magnified plot in the inset, the stray light has a small influence on the high-energy wavelength band. Then, we must address the problem that the low-energy wavelength band with the strong absorption is influenced more easily by the stray light than the high-energy wavelength band.

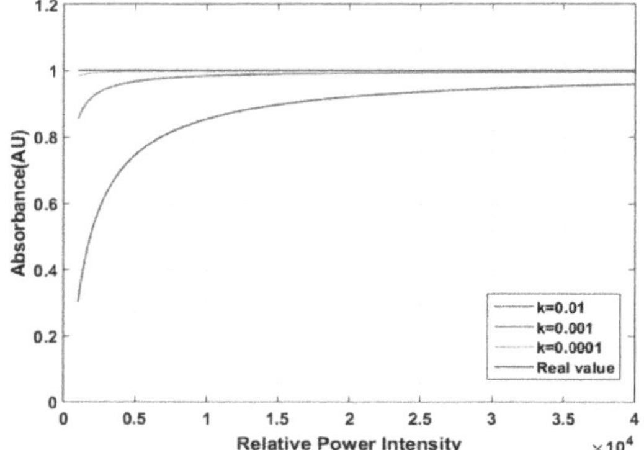

Figure 5. Influence of stray light on the low-energy wavelength band.

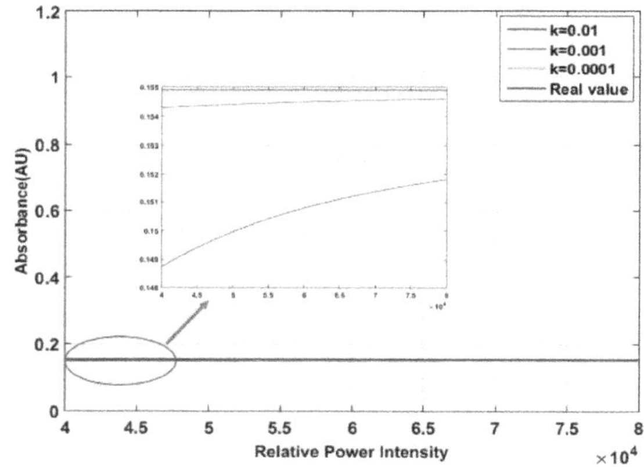

Figure 6. Influence of stray light on the high-energy wavelength Band.

The simulation indicates that the stray light has a great influence on the low-energy wavelength band. In the actual measurement, the experimental conditions are more complicated. The intensity of stray light produced by each sampling point has a different contribution to the transmission light intensity.

3.2. *Suppression of Stray Light by the Split Waveband Scan Method*

According to the analysis results, we propose a split waveband scan method. It is similar to the method of decreasing the stray light radiation. We select a filter and place it between the light source and slit. When we choose the high-pass filter, the signals only appear at the short waveband, and at the long waveband, if some signals appear at the same time, it should be the stray light produced by the short waveband. Whereas, if we select the low-pass filter, the result will be reversed. Thus, we can obtain the entire spectrum by combining the parts of two test results which have no stray light.

This method can realize the suppression of the stray light. The switching of the filter is realized in front of DMD window by the rotor controlled by the electric system.

The absorbance curve of the 95% ethanol solution is shown in Figure 7. As the requirements, our team make two types of filters. One is short-wave pass filter whose cutoff wavelength is 2210 nm, and the transmission is higher than 95% in 1350–2200 nm. When the wavelength is greater than 2210 nm, the transmission will be 0.01%. The other is a long-wave pass filter whose cutoff wavelength is 2190 nm, and the transmission is higher than 95% in 2200–2500 nm. When the wavelength is less than 2190 nm, the transmission will be 0.01%.The first filter allows high-relative-power waveband (1350–2200 nm) to pass. The other one allows low relative-power waveband (2200–2450 nm) to pass. When putting the filters into the spectrometer, the scan waveband has been divided into two parts: a high relative-power waveband and a low relative-power waveband. At the high power waveband, there is a noticeable difference between the two curves. This is caused by the decrease of the stray light which originates from the low power band. At the low power waveband, the absorbance value tested by the new method is greater than that of the traditional method. This indicates that the method of the split waveband scan can suppress parts of the impact of the stray light and make the measurement result of the absorbance increase.

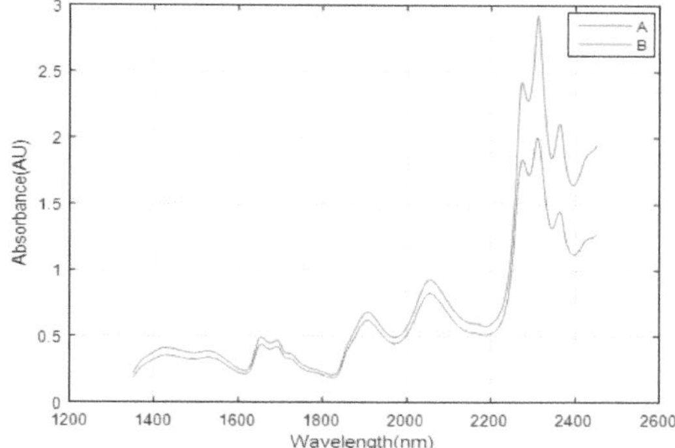

Figure 7. Absorbance spectrum of 95% ethanol. The blue line A is the absorbance curve of the test by the traditional scan method. The red line B is the absorbance curve of the test by the new scan method.

4. SNR Improvement of Low Relative-Power Waveband by a New Hadamard Mask

The NIR analysis technique is based on the small change detection in a strong background signal. The level of the SNR will have an impact on the accuracy of the analysis results, which is an important indicator [8]. In part 3, we have already provided a new scan method to suppress parts of the stray light impact, but it can't improve the SNR of the low-relative-power waveband. Thus, a new Hadamard mask of variable-width stripes is put forward to address the issue.

4.1. Impact on SNR of Different Spectral Energies

The noise source of the spectrometer includes the noise from the detector circuit and light source, which determines the SNR of the HT spectrometer [6,12–15]. Assume that n is the order of Hadamard matrix. The root mean square (RMS) of the illumination noise after the HT will be $\sqrt{\frac{2n}{n+1}}$ times higher than the original [16]. That of the detector signal noise after the HT will be $\frac{2\sqrt{n}}{n+1}$ times higher than the original [15]. Further, the SNR gain becomes $\frac{\sqrt{n}}{2}$ [16]. In [8], Xu et al. concluded that if we want to give priority to select the HT scan mode, some conditions should be satisfied. When the order n is sufficiently large so that the detector noise is equal to the illumination noise, the total noise after the HT will be lower than that after the column scan. Then, the HT scan mode will be correct.

According to [8], the SNR for the column scan method and S matrix of the HT is respectively expressed:

$$\mathrm{SNR_S} = \frac{D}{\sqrt{(D\eta)^2 + \Delta^2}} \tag{3}$$

$$\mathrm{SNR_H} = \frac{D}{\sqrt{\left(D\eta \times \sqrt{\frac{2n}{n+1}}\right)^2 + \left(\frac{2\sqrt{n}}{n+1}\Delta\right)^2}} \tag{4}$$

where D is the real spectral intensity of the light source, η is the stability of the light source, Δ is the electronic system noise collected by the analog-to digital converter (AD), and n is the matrix order of HT. Based on Equations (3) and (4), we obtain the relationship between relative power intensity and SNR, and the relationship between HT order and SNR, as shown in Figures 8 and 9.

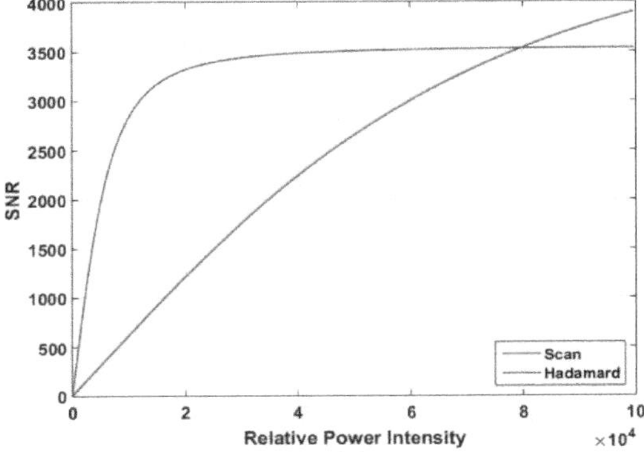

Figure 8. Signal-to-noise ratio (SNR) of two scan methods for light source energy dropping off in proportion.

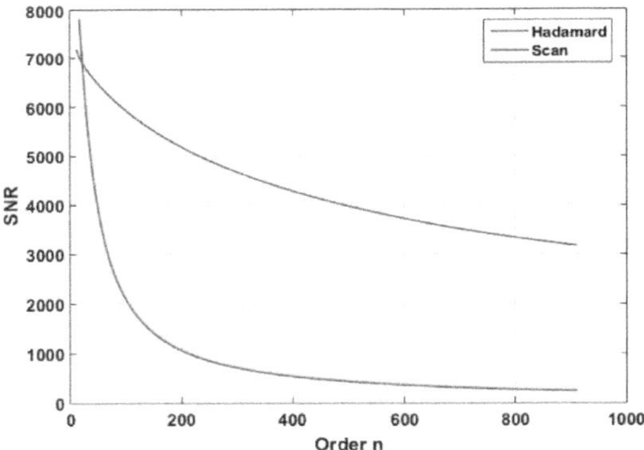

Figure 9. SNR of two scan methods for stable light source.

When the order is 107, the RMS of the light source drops off in proportion, and the requirement of the HT scan mode is satisfied. Figure 8 shows the HT scan has an advantage over the column scan method under the weak light intensity.

When the light source is stable and n = 107, the relative power intensity of the spectrum is 10000. The curve of the change in SNR following n is shown in Figure 9. When the order n increases, the SNR of the two scan methods tends to be decreased. Whereas, compared with the column scan method, the SNR of the HT scan method is decreased slowly. Because the light source energy has a different distribution in the entire waveband, the SNR of the spectrometer will be influenced. We must think of a method to change the condition of the nonuniform distribution of the spectral energy and improve the SNR of the spectral edges with low energy.

4.2. Increase in SNR of Low-Energy Waveband by New HT Scan Mask

To overcome the impact, we propose a new coding mask, a Hadamard mask of variable-width stripes (V_W-Hadamard) in the DMD HT spectrometer. By optimizing the mask, the SNR of the low-energy waveband can be improved. There are three components that determine the resolution: the slit width, the optical transfer function, and DMD resolution. Because of the small pixel size of the DMD, the DMD resolution is not the major determinant [9]. Thus, we plan to change the resolution in different patterns of the DMD. By making sacrifices on the resolution of the edge DMD patterns, the energy loss in the low-energy waveband can be reduced. The width of the scanning stripes is determined by the corresponding spectrum energy in each stripe. The higher the corresponding spectrum energy is, the narrower the stripe width is. Then, the SNR of the edge spectrum with low energy can be improved. Xu et al. proposed a HT scan method by using a mask of variable-height stripes [8]. This method greatly improved the SNR and the even distribution of light source spectral energy. However, because of the change in the height of the stripes, the DMD pattern could not be utilized well. Moreover, a part of the spectral energy was lost. In this study, the V_W-Hadamard mask can make full use of the spectral energy because of the adaptive variation in the width of the stripes. The new HT mask has an advantage on the energy utilization over the traditional mask. The flow chart of the mask production is outlined in Figure 10.

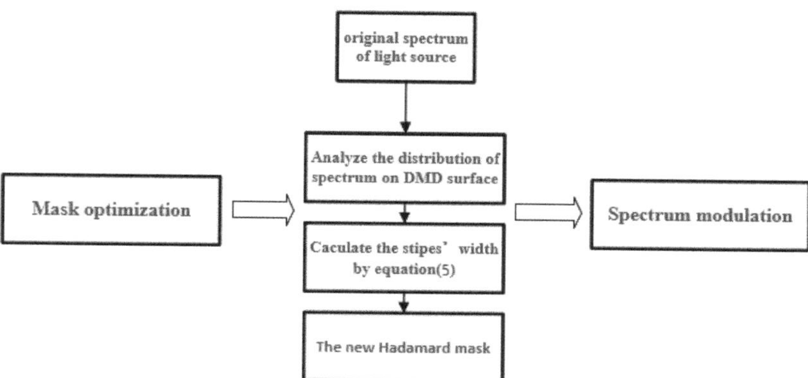

Figure 10. Flowchart of mask optimization.

Assume E_m is the median power intensity, and W_m is the strip width of E_m, which is taken as the width of three columns of the micromirrors. The width W of the other changing strip corresponding to the power intensity of E can be calculated by

$$\frac{E_m}{E} = \frac{W}{W_m} \tag{5}$$

Whatever the result of the equation is, there exists a maximum wavelength which makes the spectral waveform meaningful. The spectral resolution should not go beyond that. Figure 11a,b separately show that the traditional and the new Hadamard mask. They are produced by the same order of 107.

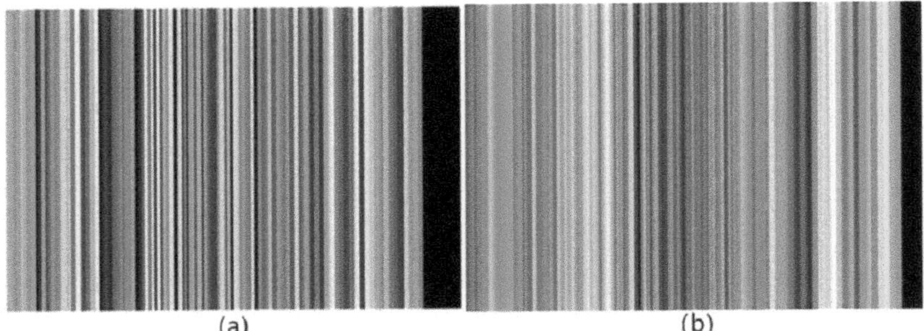

Figure 11. Hadamard masks: (**a**) traditional Hadamard mask (**b**) new Hadamard mask. They are produced by the same order of 107.

In order to verify the performance improvement of the HT spectrometer based on the new coding mask, we perform some tests on the light source spectrum. The DMD resolution is 912 × 1104. There are 912 columns of micromirrors in the direction of the spectrum dispersion. The curves of the relative power intensity acquired by the column scan method with different sampling points are presented in Figure 12. With the increase in the number of sampling points, the spectral energy received by the detector decreases. Furthermore, the distribution of each wavelength corresponding to the energy is more even than before. Considering the curves in Figures 9 and 12, the number of sampling points can't be too large. Thus, we choose the orders of 107 and 227 to perform the test and analysis. The spectrum is divided into two parts: a high-energy wavelength band (1350–2170 nm) and low-energy wavelength band (2170–2450 nm). Based on reference 8, the SNR_{vi} by the new mask scanning method can be calculated by

$$SNR_{Vi} = \frac{D_i}{\sqrt{\left(D_i\eta \times \sqrt{\frac{2N_i}{N_i+1}}\right)^2 + \left(\frac{2\sqrt{N_i}}{N_i+1}\Delta\right)^2}} \quad (6)$$

where i denotes the i-th scanning stripe, SNR_{vi} is the SNR of the i-th stripe region, D_i is the spectral intensity of the i-th stripe scanning region, and N_i is the matrix order of the HT in the i-th scanning stripe. The curves of the spectral SNR calculated by the equations (Equations (3), (4) and (6)) are shown in Figures 13 and 14.

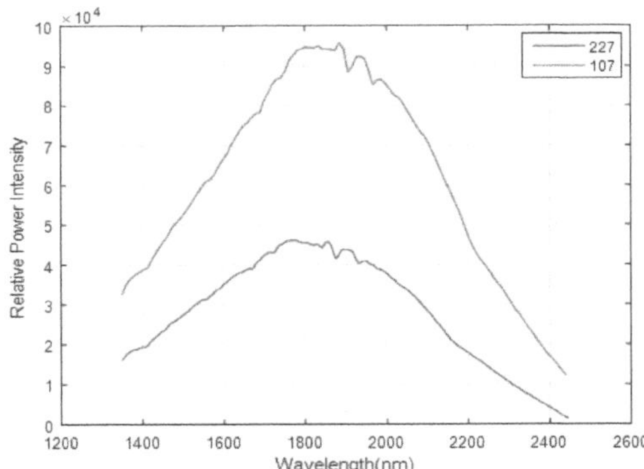

Figure 12. Relative power intensity of the light source. The curves are obtained by the column scan mode with the orders of 107 and 227.

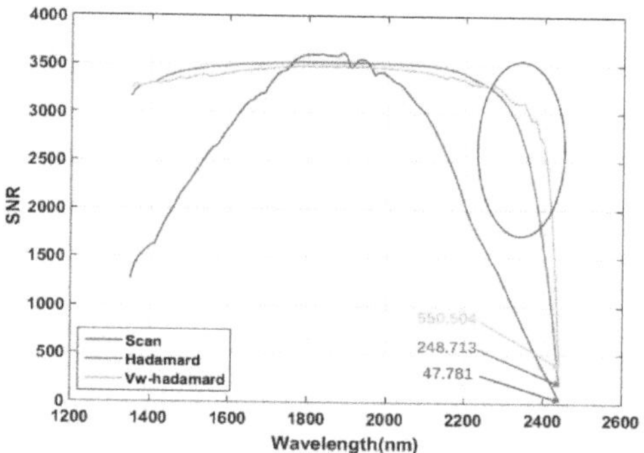

Figure 13. SNR of the spectrometer with the order of 107.

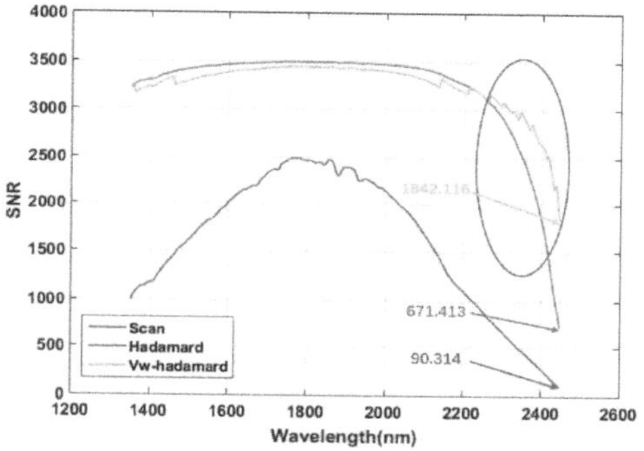

Figure 14. SNR of the spectrometer with the order of 227.

For n = 107, the average SNR obtained by the V_W-Hadamard method (SNR_V) in the short-wave sideband is 3289.153, which is 2.034 times that by the scanning spectrometer (SNR_S). That by the traditional Hadamard method (SNR_H) is 3293.121, which is 2.036 times that of SNR_S. For the average SNR in the long-wave sideband (2220–2450 nm), SNR_V is 2.972 times that of SNR_S and 1.062 times that of SNR_H. For the minimum SNR in the short-wave sideband, SNR_V is 2.547 times that of SNR_S and 1.026 times that of SNR_H. For the minimum SNR in the long-wave sideband, SNR_V is 11.52 times that of SNR_S and 2.213 times that of SNR_H. When n = 227, the average SNR of the V_W-Hadamard spectrometer is improved compared to that of the Hadamard. Particularly, for the minimum SNR in the long-wave sideband, SNR_V is 20.397 times that of SNR_S and 2.743 times that of SNR_H. According to the results, the V_W-Hadamard mask with the order of 227 performs better than that with the order of 107 in the long-wave sideband.

The curves of the DMD resolution by the two methods are presented in Figure 15. Because the DMD resolution matches with the spectral resolution, the spectral resolution (the yellow curve) in Figure 15 is the same as the DMD resolution of the traditional Hadamard. From Figure 15, we can see that the DMD resolution has a slight increase in 1450–2220 nm for the new Hadamard when the order is 107 and 227. However, the resolution increase is not useful because the DMD resolution should match with the spectral resolution. This means if the spectral resolution is lower than the DMD resolution, the resolution of the spectrometer will depend on spectral resolution. In 1300–1450 and 2220–2450 nm, the DMD resolution of the new Hadamard is decreased. By combining SNR with the DMD resolution, we conclude the SNR have been improved by means of sacrificing the resolution in the two sideband.

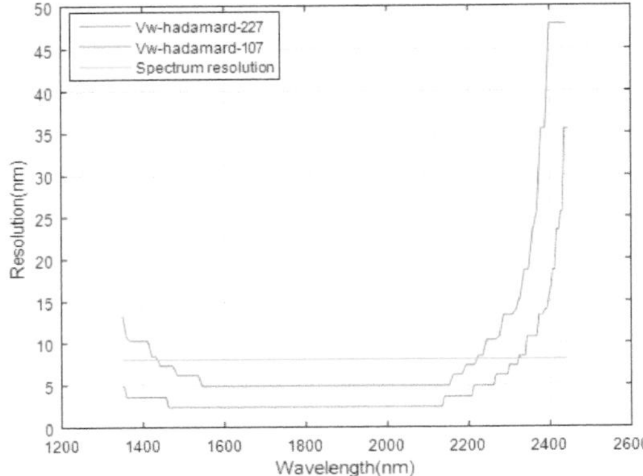

Figure 15. Digital micromirror device (DMD) resolution of the new Hadamard and the traditional Hadamard. The blue and red curves are obtained by the V_W-Hadamard mask. The yellow curve is the spectral resolution which is the same as the DMD resolution of the traditional Hadamard.

5. Conclusions

In HT spectrometers with a DMD, we analyze the influence of stray light. It is shown that the stray light mainly affects the spectral energy distribution of the light source. It can reduce the measurement value of the spectral absorbance and make the SNR inconsistent. We address these problems from two aspects. One is to mitigate the impact of the stray light by the double filter strategy, and the other is to improve the SNR by a new Hadamard mask. In addition, the experiments and simulations are conducted and compared between the new scanning method and traditional HT scanning method. The SNRs of the spectrometers are compared for different methods when the order is 107 and 227. The results demonstrate that the new method and mask can effectively suppress the stray light and make the distribution of the spectral energy more uniform. Meanwhile, the SNR also is improved.

Author Contributions: Conceptualization: H.L. and J.L.; methodology: J.X.; software: J.X.; validation: Z.L., H.L. and J.L.; formal analysis: Z.L.; investigation: J.Z.; resources: H.L.; data curation: J.Z.; writing—original draft preparation: Z.L.; writing—review and editing: Z.L.; visualization: J.L.; supervision: H.L.; project administration: H.L.; funding acquisition: H.L.

Funding: The work described in this paper is supported by the National Natural Science Foundation of China under grant No. 61875036, Projects of Science and Technology Development Plan of Jilin Province (20190302049GX), the State Key Laboratory of Applied Optics, and the Lab of Space Optoelectronic Measurement & Perception (LabSOMP-2018-05).

Acknowledgments: We acknowledge financial support by the National Natural Science Foundation of China under grant No. 61875036, Projects of Science and Technology Development Plan of Jilin Province (20190302049GX), the State Key Laboratory of Applied Optics, and the Lab of Space Optoelectronic Measurement & Perception (LabSOMP-2018-05).

Conflicts of Interest: The authors declare no conflict of interest.

References

1. Harwit, M.; Sloane, N.J.A. *Hadamard Transform Optics*, 1st ed.; Academic Press: San Diego, CA, USA, 1979; pp. 1–264.
2. Hirschfeld, T.; Wyntjes, G. Fourier transform vs. Hadamard transform spectroscopy. *Appl. Opt.* **1973**, *12*, 2876–2880. [CrossRef] [PubMed]
3. Kearney, K.J.; Ninkov, Z. Characterization of a digital micromirror device for use as an optical mask in imaging and spectroscopy. In Proceedings of the SPIE 3292 on Spatial Light Modulators, San Jose, CA, USA, 20 April 1998; pp. 81–92.

4. Wang, X.-D.; Liu, H.; Lu, Z.-W.; Song, L.-W.; Wang, T.-S.; Dang, B.-S.; Quan, X.-Q.; Li, Y.-P. Design of a spectrum-folded Hadamard transform spectrometer in near-infrared band. *Opt. Commun.* **2014**, *333*, 80–83. [CrossRef]
5. Wang, X.; Liu, H.; Juschkin, L.; Li, Y.; Xu, J.; Quan, X.; Lu, Z. Freeform lens collimating spectrum-folded Hadamard transform near-infrared spectrometer. *Opt. Commun.* **2016**, *380*, 161–167. [CrossRef]
6. Zhang, W.; Zhang, Z.; Gao, L. Study of using complementary S matrix to enhance SNR in Hadamard spectrometer. *Optik* **2014**, *25*, 1124–1127. [CrossRef]
7. Quan, X.; Liu, H.; Lu, Z.; Chen, X.; Wang, X.; Xu, J.; Gao, Q. Correction and analysis of noise in Hadamard transform spectrometer with digital micro-mirror device and double sub-gratings. *Opt. Commun.* **2016**, *359*, 95–101. [CrossRef]
8. Xu, J.; Liu, H.; Lin, C.; Sun, Q. SNR analysis and Hadamard mask modification of DMD Hadamard transform near-infrared spectrometer. *Opt. Commun.* **2017**, *383*, 250–254. [CrossRef]
9. Texas Instruments, DMD® Spectrometer Design Considerations. Available online: http://www.ti.com.cn/cn/lit/an/dlpa049/dlpa049.pdf (accessed on 1 August 2014).
10. Wang, Y.; Liu, H.; Li, J.H.; Lu, Z.F.; Xu, J.L.; Chen, B. Research on near-infrared spectrometer based on DMD. *Infrared and laser engineering*. (accepted).
11. Zong, Y.; Brown, S.W.; Johnson, B.C.; Lykke, K.R.; Ohno, Y. Correction of stray light in spectrographs: Implications for remote sensing. In Proceedings of the SPIE 5882 on Earth Observing Systems X, San Diego, CA, USA, 22 August 2005; pp. 588201–588208.
12. Wuttig, A.; Riesenberg, R. Sensitive Hadamard transform imaging spectrometer with a simple MEMS. In Proceedings of the SPIE 4881 on Sensors, Systems, and Next-Generation Satellites VI, Crete, Greece, 8 April 2003; pp. 167–178.
13. Nitzsche, G.; Riesenberg, R. Noise, fluctuation, and Hadamard-transform spectrometry. In Proceedings of the SPIE 5111 on Fluctuations and Noise in Photonics and Quantum Optics, Santa Fe, NM, USA, 16 May 2003; pp. 273–282.
14. Ye, M.; Ye, H.N.; Yang, X.; Wang, H.; Wang, R. The limited source in Hadamard transform optics. In Proceedings of the SPIE 7130 on Precision Mechanical Measurements, Anhui, China, 31 December 2008; p. 71303X.
15. Mende, S.B.; Claflin, E.S.; Rairden, R.L.; Swenson, G.R. Hadamard spectroscopy with a two-dimensional detecting array. *Appl. Opt.* **1993**, *32*, 7095–7105. [CrossRef] [PubMed]
16. Xiang, D.; Arnold, M.A. Solid-state digital micro-mirror array spectrometer for Hadamard transform measurements of glucose and lactate in aqueous solutions. *Appl. Spectrosc.* **2011**, *65*, 1170–1180. [CrossRef] [PubMed]

 micromachines

Article

Measuring Ocular Aberrations Sequentially Using a Digital Micromirror Device

Alessandra Carmichael Martins * and Brian Vohnsen

Advanced Optical Imaging Group, School of Physics, University College Dublin, Dublin D04, Ireland; brian.vohnsen@ucd.ie

* Correspondence: alessandra.carmichaelmartins@ucd.ie; Tel.: +353-1-716-2352

Received: 5 January 2019; Accepted: 8 February 2019; Published: 12 February 2019

Abstract: The Hartmann–Shack wavefront sensor is widely used to measure aberrations in both astronomy and ophthalmology. Yet, the dynamic range of the sensor is limited by cross-talk between adjacent lenslets. In this study, we explore ocular aberration measurements with a recently-proposed variant of the sensor that makes use of a digital micromirror device for sequential aperture scanning of the pupil, thereby avoiding the use of a lenslet array. We report on results with the sensor using two different detectors, a lateral position sensor and a charge-coupled device (CCD) scientific camera, and explore the pros and cons of both. Wavefront measurements of a highly aberrated artificial eye and of five real eyes, including a highly myopic subject, are demonstrated, and the role of pupil sampling density, CCD pixel binning, and scanning speed are explored. We find that the lateral position sensor is mostly suited for high-power applications, whereas the CCD camera with pixel binning performs consistently well both with the artificial eye and for real-eye measurements, and can outperform a commonly-used wavefront sensor with highly aberrated wavefronts.

Keywords: wavefront sensing; digital micromirror device; ocular aberrations

1. Introduction

Quantification of aberrations is important in a number of applications and needed to optimize performance as, for example, with adaptive optics. One area of special relevance is ophthalmology, where the optical quality of the human eye is important not only for acute vision but also for diagnostic retinal imaging applications.

The optical performance of the human eye is limited not by diffraction but rather by the amount of monochromatic and chromatic aberrations. Ocular wavefront aberrations are mainly induced by the cornea and the crystalline lens and become increasingly important with a larger pupil size. The wavefront aberrations commonly refer to the conjugate pupil plane of the eye, where deviations from a planar wavefront refer to aberrations that prevent light from being focused onto a diffraction-limited spot on the retina. The aberrations for monochromatic light are expressed as a linear combination of orthonormal circular Zernike polynomials weighted by a series of Zernike coefficients, expressed in either μm or wavelength units [1].

Many methods have been proposed since the 1960s to measure aberrations of the human eye both objectively and subjectively including modified aberrometers [2,3], ray tracing techniques [4], and indirectly via retinal images [5]. However, Hartmann–Shack wavefront sensors (HS-WFSs) have become the preferred devices to measure the wavefront and intensity distribution of backscattered light from the retina [6]. The HS-WFS uses a tightly focused beacon of light in the retinal plane, which serves as secondary point source for the wavefront sensing, with a lenslet array that samples the local distribution of wavefront tilt in the pupil plane. Complementary metal-oxide-semiconductor (CMOS)-based HS-WFS can capture aberration changes at 100′s of Hz, although common CCD-based

HS-WFS are limited to 10's of Hz. In either case, cross-talk between adjacent lenslets limit the dynamic range. Modifications to HS-WFS have been proposed to improve the dynamic range by using a liquid-filled lenslet array [7] and by replacing it with a liquid crystal display [8,9], although only few have been tested with ocular aberrations.

As the need to determine ocular aberrations has become increasingly important for personalized refractive corrections using custom LASIK [10], and for the understanding of a number of refractive problems ranging from keratoconus [11] to increased high myopia [12], establishing a wavefront sensing technique that provides a high dynamic range, resolution, and high speed suitable for the human eye is crucial. Digital micromirror devices (DMDs) can operate at speeds in the 10's of kHz range, while sequential scanning provides a large dynamic range due to the lack of a lenslet array; therefore, avoiding the appearance of cross-talk, as recently proposed by the authors in a DMD-WFS [13]. Additionally, DMDs have recently been used for ophthalmic applications in retinal imaging [14] and psychophysical measurements [15]. A somewhat related technique uses scanning of an incident beam of light in the pupil plane to capture multiple retinal images [16]. Here, the narrow beam of light is incident near the pupil center and scanning is only done for light exiting the eye by sequential aperture scanning of the pupil, analogous to the parallel sampling by a lenslet array in the HS-WFS. Centroiding methods of the imaged point-spread-function (PSF) for ocular aberrations have been compared [17] and can be tuned to provide more accurate wavefront reconstructions in the presence of noise.

Here, the wavefront sensing technique using a DMD is used to measure ocular aberrations of the human eye by performing sequential zonal scanning of the wavefront. The experimental setup and method are explained in Section 2 using two types of detectors. Results are shown in Section 3.1 for an artificial eye and in Section 3.2 for human eyes. A discussion about the results and technique can be found in Section 4, followed by the conclusion of the study in Section 5.

2. Materials and Methods

A wavefront sensor based on sequential scanning of a reflective cell with a DMD (V-7001 VIS, Vialux, Chemnitz, Germany) is used to measure ocular aberrations. The system, which is described in detail in [13], has been adapted for real-time aberration sensing of the human eye. A schematic of the setup can be seen in Figure 1. Essentially, a narrow near-infrared (850 nm) beam of 200 μW entering the eye is focused onto the retina to create a secondary point source suitable for wavefront sensing. Backscattered light exiting the eye is truncated by a 4 mm iris and the pupil is imaged by a 4f telescope onto the DMD. The DMD is comprised of 1024 × 768 mirrored square pixels of 13.7 μm that allow binary positioning at ±12° (and optical angles of ±24°) at up to 22.7 kHz. The DMD divides the imaged pupil into equal-sized cells, consisting of a square array of micromirrors, which are sequentially activated, and the reflected light is focused onto a position detector. Due to diffraction effects from the DMD, an iris is used in a conjugate retinal plane to only allow the pass of the 0th order. In this study, two different position detectors are used:

(1) A 2D-lateral resistive position detector (PDP90A, Thorlabs™, Newton, NJ, USA) of up to 0.75 μm spatial resolution, was used to register the central position of the PSF centroid (x,y) for each activated DMD cell. Although its angular resolution suffices, the limited sensitivity of this device prevents it from being used to determine ocular aberrations, but it is included here as a proof-of-principle in Section 3.1.1. Indeed, it may well find applications where power limitations are of less concern, such as in the characterization of laser beams.

(2) A CCD camera (Scientific Camera 1501M-USB, Thorlabs™, Newton, NJ, USA) with 6.45 μm pixel pitch and 14-bit digital output is used for the rest of the results in Sections 3.1.2 and 3.2, allowing for high brightness variations and binning of pixels to increase acquisition speed when acquiring images of the PSF from which the centroid position can be determined.

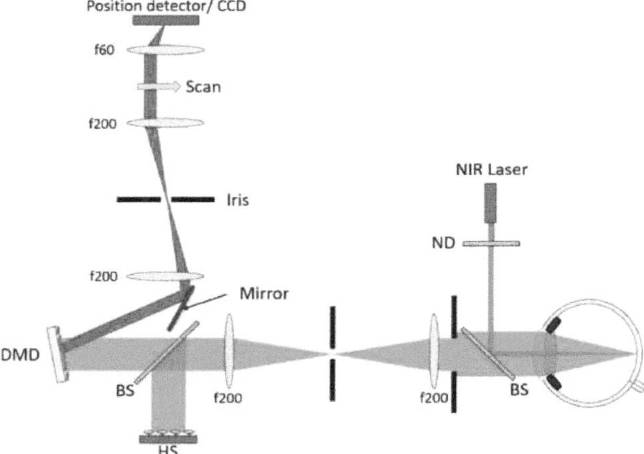

Figure 1. Schematic of the system used to measure ocular aberrations using a digital micromirror devices (DMD). A 4 mm beam of near-infrared (IR) light backscattered from the retina is sequentially scanned in the pupil plane by the DMD, which directs light onto a position detector via a plane mirror to capture the maximum light. The detector captures the (x,y) centroid coordinates of the point-spread-function (PSF) in the case of the lateral position detector, or captures images of the PSF in case of the charge-coupled device (CCD) camera, for each scanned section of the light. All lenses used are antireflection-coated achromatic doublets.

The DMD, the position detector, and the CCD are all programmed and synchronized via LabVIEW™. Thus, the position detector captures the PSF centroid coordinates at a speed of 1.5 kHz, whereas the CCD captures the PSF images at a speed of 13 frames-per-second (FPS) without pixel binning and 1 ms exposure, but faster with pixel binning of up to 24 × 24 pixels. For example, with 10 × 10 pixel-binning the speed limit is 77 FPS. For both sensors, the aberration is determined with respect to that of a plane reference wave obtained by placing a flat mirror in the pupil plane. A conventional Hartmann–Shack wavefront sensor (HS-WFS 150-5 C, 73 × 73 lenslets, Thorlabs™, Newton, NJ, USA) is placed in a conjugate pupil plane and is used for comparison and verification purposes.

First, a test was performed with an artificial eye comprised of a thin ophthalmic trial lens adjacent to a flat mirror placed in the pupil plane. The DMD-WFS scan was implemented with two sampling densities: (a) 5 × 5 DMD cells of 800 μm (58 × 58 pixels) obtaining the corresponding 25 PSF images, and (b) 10 × 10 DMD cells of 400 μm (29 × 29 pixels) corresponding to 100 PSF images. For each sampling density of the DMD, the PSF images were acquired using 1 × 1, 2 × 2, 4 × 4, and 8 × 8-pixel-binning of the CCD camera, recovering images of 1392 × 1040, 696 × 520, 348 × 260, and 174 × 130 pixels, respectively. A plane reference wave was also attained for each case, and the effects of binning compared.

Secondly, ocular aberration measurements were performed in the right eye of 4 emmetropic subjects (equivalent sphere between 0 and −1 diopters, as measured with an EyeNetra™ autorefractor) and one myopic subject (with an equivalent sphere of −7D) whose pupils were dilated, and accommodation partially paralyzed, with two drops of 1% tropicamide. To ensure a point source on the retina of myopic subjects, a corrective trial lens is placed in a prior conjugate plane to the pupil, making use of a one-to-one 4f system composed of two 150 mm focal length achromatic lenses. A bite bar was used to ensure good centration of the eye's pupil and reduce head movement throughout the measurements. For their own commodity, subjects were asked to take a short break between one scan measurement and the next, hence slightly readjusting their position for each measurement.

For both cases, the Zernike wavefront parameters obtained from the HS-WFS were recorded up to the 4th radial order. Subsequently, the DMD-WFS scan was performed at 13 FPS, adding up to a total time of 2 s for a complete 5 × 5 sampling and 8 s for a 10 × 10 scan. The PSF images were post-analyzed in Matlab^{TM} to determine the center-of-gravity centroid translations in cartesian coordinates between the aberrated wavefront and that of the reference for each DMD cell. These translations were used to calculate the Zernike coefficients from which the least-square wavefront reconstruction was performed. The procedure of this method is explained in detail in [13]. For 5 × 5 DMD sampling, the PSF corresponding to DMD cells illuminated across more than 50% of their area were analyzed and used for the reconstruction, while for 10 × 10 sampling, only those 100% illuminated were considered. To further examine the effects of using 50% illuminated area versus 100% illuminated area of DMD cells and that of binning, the reconstruction procedure here followed for 5 × 5 DMD cell sampling was compared in Figure 2 for a simulated wavefront with astigmatism, coma, and trefoil, similar to that from [18] for a typical human eye. Reconstructing the wavefront only using the fully illuminated DMD cells, provides more accurate results for a static wavefront. However, when sampling a dynamic wavefront, such as that of the real eye, with a low density and only considering cells illuminated in their entirety, could be insufficient for accurate measurements. On the other hand, increasing pixel binning in the detection camera allows for faster acquisition and is seen to maintain accuracy up to 8 × 8 binned pixels (deviation <4%), after which the accuracy of the wavefront reconstruction tends to fall.

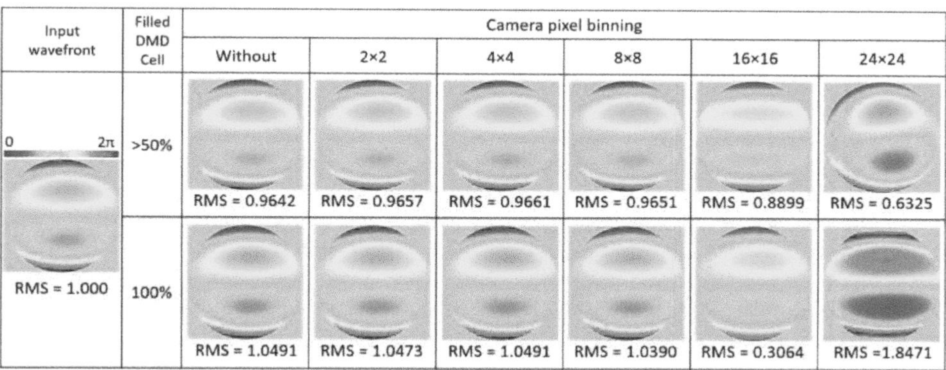

Figure 2. Comparison of wavefront reconstructions of a simulated wavefront with astigmatism, coma, and trefoil. First, using DMD cells which are at least 50% illuminated with a 4 mm beam, and second, those 100% illuminated by the incoming beam. For each case, the effect of square pixel binning of ×2, ×4, ×8, ×16, and ×24 pixels in the detector camera is compared. All root-mean-square (RMS) values are given in μm.

3. Experimental Results

Wavefront diagrams and root-mean-square (RMS) wavefront values shown in this section are exempt from tip and tilt, as these are not representative of the ocular aberrations as such, and a minimal decentration is necessary to avoid corneal reflections in the HS-WFS measurements.

3.1. Wavefront Aberrations with an Artificial Eye

3.1.1. Lateral Position Detector

The lateral position detector provides the cartesian coordinates of an incident beam with respect to a predetermined center point and is commonly used for system alignment. Here, it was used to determine the centroid coordinates of the PSF and displacements were given with respect to that of a focused plane reference wave. The reconstruction of aberrated wavefronts induced by four different trial lenses can be seen in Figure 3 and compared to the wavefront measured by the HS-WFS. Astigmatism was achieved with two crossed positive and negative power cylindrical lenses. All measurements shown were performed with 5 × 5 DMD cell sampling. Increasing the sampling density to 10 × 10 decreased the amount of optical power per cell that reached the position detector, and; therefore, limited its accuracy with the available power.

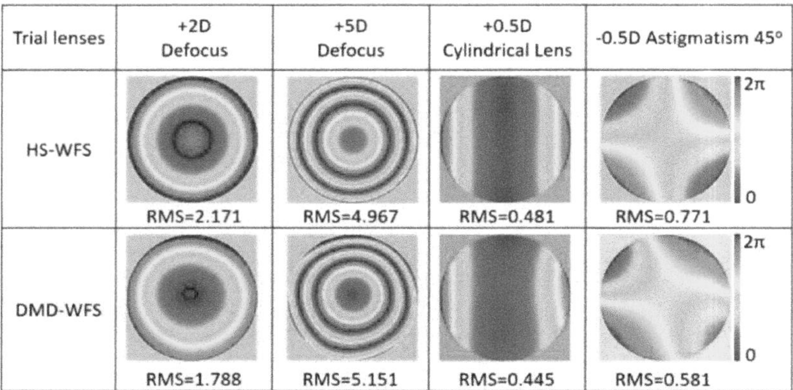

Figure 3. Wavefront reconstructions of an aberrated wave across a 4 mm pupil with ophthalmic trial lenses using the 5 × 5 sampling DMD-WFS method and a 2D-lateral position detector placed in the image plane. All RMS values are given in μm.

3.1.2. CCD Camera Detector

The wavefront reconstructions of the artificial eye using the Zernike coefficients measured by the HS-WFS and the DMD-WFS were included in Figure 4 for a 4 mm pupil and two sets of ophthalmic lenses: a) defocus with a +8D spherical lens, b) astigmatism with two crossed −4D and +4D cylindrical lenses, and c) combination of defocus and astigmatism with a +8D spherical lens and crossed −4D and +4D cylindrical lenses. Given that the trial lenses were used in conjunction with a flat mirror, the double pass of the beam caused induced aberration to lay beyond the range of the HS-WFS. Different binning options ranging from one to eight pixels were included when measured with the DMD-WFS. The RMS values of the Zernike coefficients were used to quantify and compare the obtained results, with only up to 2.5% deviation for different pixel-binning options. However, a difference of up to 30% was observed when compared to the HS-WFS, where the aberrations were underestimated. The obtained data comparing both methods in terms of Zernike coefficients as well as sphere and cylinder power is detailed in Table 1.

Table 1. Data comparison between the Zernike coefficients $c_{2,0}$ and $c_{2,2}$ and the equivalent sphere and cylinder power for a 4 mm pupil, with an artificial eye obtained by the HS-WFS and the DMD-WFS without binning.

	HS-WFS				DMD-WFS			
	$c_{2,0}$[μm]	$c_{2,2}$[μm]	Sphere[D]	Cyl.[D]	$c_{2,0}$[μm]	$c_{2,2}$[μm]	Sphere[D]	Cyl.[D]
a)	6.638	0.308	11.50	0.75	7.519	−0.024	13.02	0.00
b)	0.627	−4.867	1.08	−11.90	0.561	−6.511	0.97	−15.95
c)	6.270	−2.472	10.85	−6.05	7.084	−3.769	12.27	−9.23

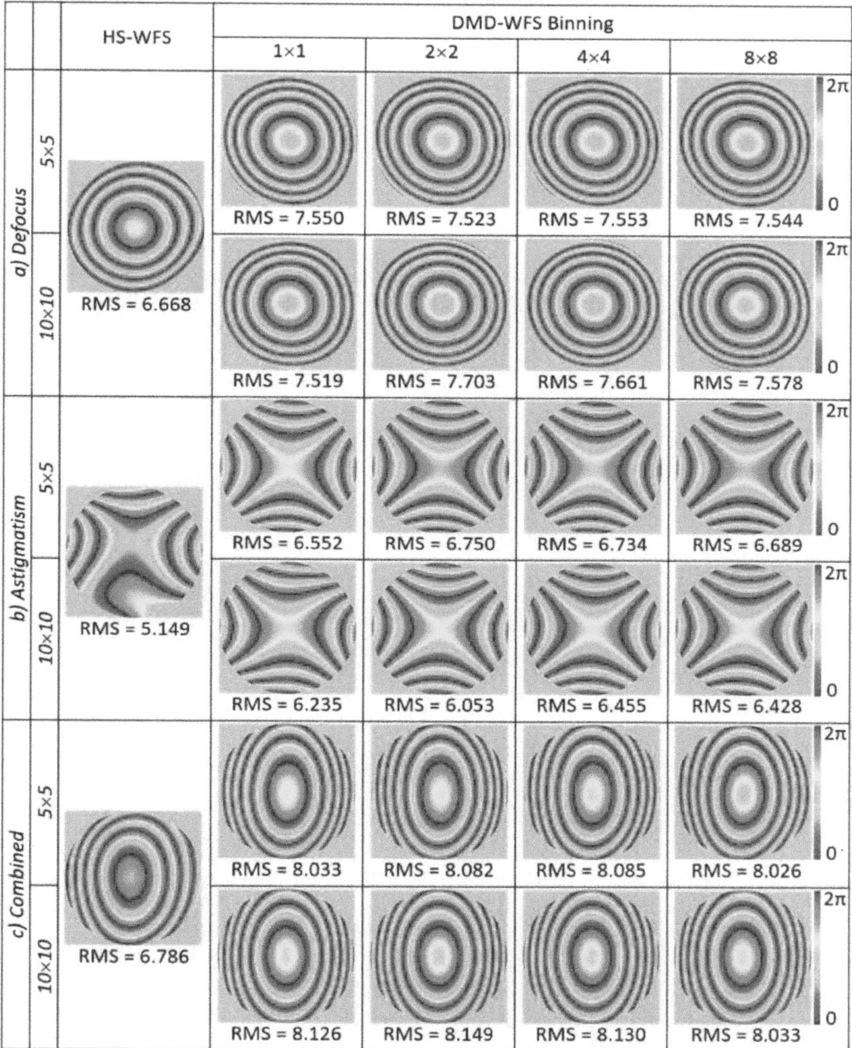

Figure 4. Comparison of defocus and astigmatic wavefront reconstructions for a 4 mm pupil with an artificial eye through Zernike coefficients between HS wavefront sensing, and for the DMD wavefront sensing method with four different binning options in the CCD camera, acquiring the PFS images for each DMD cell. All RMS values of the Zernike coefficients are given in μm.

3.2. *Wavefront Aberrations of the Real Eye*

Measurements of ocular aberrations of five healthy subjects, four of which were emmetropes and one which was myopic (−7D), are shown in Figure 5 for 5 × 5 sampling density and in Figure 6 for 10 × 10 sampling density. In both cases, four types of pixel binning in the CCD camera were included. Wavefronts were quantified using the RMS value of the Zernike coefficients, given in μm.

Larger variations between the reconstructed wavefronts appeared when measuring the real eye in comparison to the previous results with the artificial eye. This could be due to the natural movement of the eye, which includes both voluntary and involuntary movements, even when fixating on a given target [19]. Fixation time lasted approximately 200–300 ms [20,21], with large variability between subjects, which falls well below the required DMD-WFS scan time. Increasing the pixel binning allowed for a higher acquisition speed to limit variations during ocular aberration measurements. Variations of 5% to 30% in the RMS values were noted between different binning options performed at the same speed for 5 × 5 DMD sampling. Denser sampling of 10 × 10 DMD cells involved higher acquisition time, causing deviations to increase between 7% and 40%.

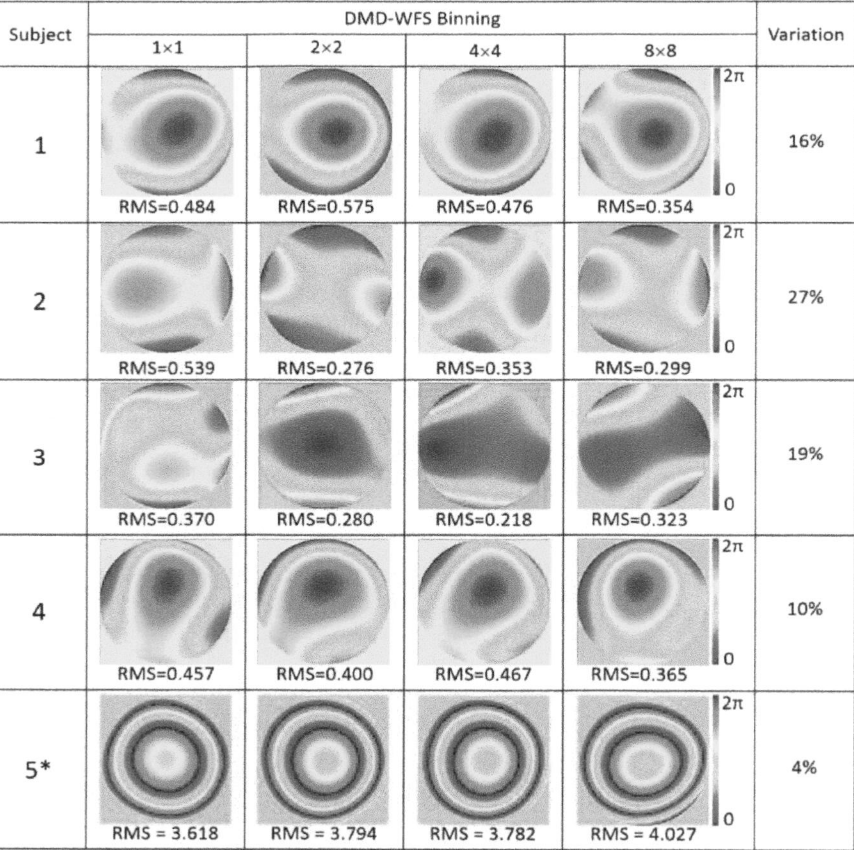

Figure 5. Ocular aberrations measured with the DMD-WFS with 5 × 5 sampling density for four subjects with normal vision and one myopic subject (−7D), marked with an asterisk (*), using DMD cell with at least 50% of its area illuminated. Quantification of wavefronts are presented as RMS values given in μm.

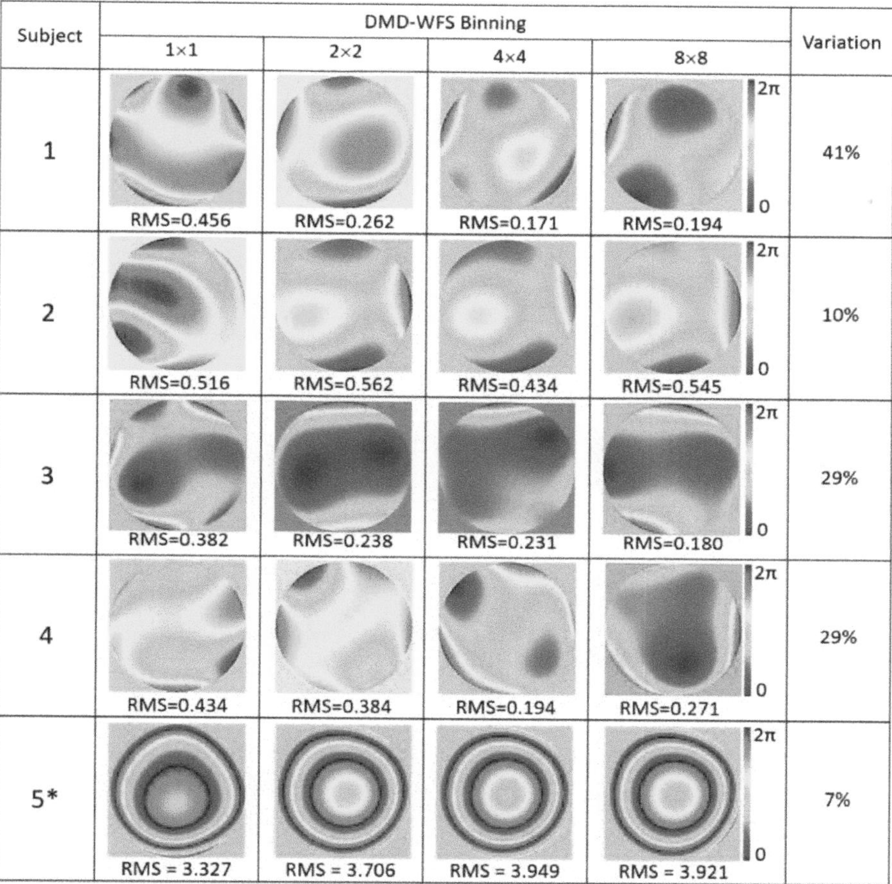

Figure 6. Ocular aberrations measured with the DMD-WFS with 10 × 10 sampling density using only 100% illuminated DMD cells for the same five subjects. The asterisk (*) denotes the myopic subject. All RMS values are given in μm.

In order to analyze the effect of scanning at different speeds, measurements were performed in the right eyes of the authors at 5, 10, 15, and 20 frames per second with both 5 × 5 and 10 × 10 DMD cell sampling. Results are included in Figure 7. For each subject, larger differences were seen in the case of 10 × 10 sampling, as the measurement time varied considerably between being performed at 5 FPS or 20 FPS.

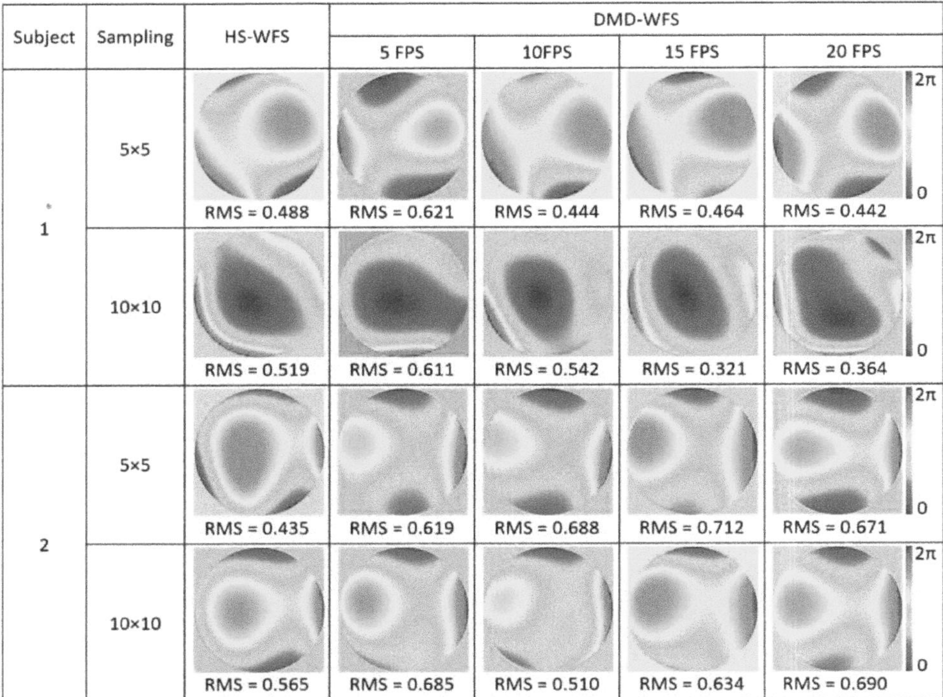

Figure 7. Comparison of the DMD-WFS scan performed at 5, 10, 15 and 20 FPS for two subjects with both 5 × 5 and 10 × 10 DMD cell sampling densities. All RMS values are given in µm.

4. Discussion

The use of a DMD to sequentially scan an aberrated wavefront using a single achromatic lens to focus light onto a position-sensitive detector allows for wavefront measurements with high dynamic range by avoiding crosstalk, which potentially limits the performance of a conventional HS-WFS. This sees applicability in ophthalmology due to the fast-increasing rates of high myopia [22]. Corneal reflections, or undesired corneal areas can also be eliminated by deactivating the corresponding cells in the DMD, such that these do not induce noise into the wavefront reconstruction. An example where reflections can cause problems can be seen in the lower part of Figure 4b for the HS-WFS wavefront with astigmatism.

The use of a position detector device provides direct determination of centroid coordinates with an accuracy of up to 0.75 µm when output voltage is maximized, avoiding the need to save large amounts of image data for each measurement, specifically in the case of large sampling densities and no pixel binning. However, given its high-power requirements for accurate detection, it did not prove feasible for ophthalmic applications.

The CCD camera with a 14-bit optical output working synchronously with the DMD was shown to measure static wavefronts with high precision in the artificial eye. Furthermore, the possibility of binning pixels in x- and y-directions allow for increased speed and lower amounts of stored data without compromising accuracy. Slight variations in the wavefront RMS values seen in Figure 4 with different binning options of up to 8 × 8 pixels are in the same order of magnitude as predicted by the simulated wavefront from Figure 2.

For ophthalmic applications, the high speed of the DMD and CCD camera to perform a complete scan is crucial. Scans performed at 13 FPS were to suffice when measuring ocular aberrations with both 5 × 5 and 10 × 10 DMD cell sampling, and the effect of scanning speed between 5 and 20 FPS was analyzed. However, variations in the measured wavefronts are still present due the continuous involuntary movements of the eye during fixation including tremors, drifts, and microsaccades [19–21], where further increasing the speed would improve accuracy and repeatability. Changes in the tear film, which is known to be dynamic, may also cause changes in the measurements of ocular aberrations [23]. Binning larger amounts of pixels in the detection camera, can help improve the signal; however, compromises accuracy above 8 × 8 pixels, as seen in Figure 2. Additionally, using slightly better centroiding methods [17] could improve accuracy. The use of a high-speed CMOS camera could potentially allow for higher speed, but ultimately a different methodology, such as single-pixel sensing, may be required to gain the upmost in terms of kHz speed [24].

5. Conclusions

Sequential scanning of a wavefront using a DMD and a single achromatic lens to measure aberrations, with near-infrared light removing the conventional lenslet array found in HS-WFS, has been achieved. This provides high dynamical range with a trade-off between high sampling densities and speed, where the latter can be increased by pixel binning. This technique grants high adaptability to different applications and is here tested to measure ocular aberrations in artificial and real eyes. Two different wavefront sampling densities were compared, and the effect of pixel binning was analyzed and found to allow for accurate reconstructions for a static artificial eye, but were subject to variations in the real eye, where increased speed would still be paramount for higher accuracy.

Author Contributions: A.C.M. performed the labview coding, experimental analysis and wrote the draft of the paper; B.V. provided matlab coding for wavefront reconstruction and edits to the draft. A.C.M. and B.V. contributed equally to the experimental design.

Funding: This research was funded by H2020 ITN MyFUN, grant number 675137.

Acknowledgments: The authors would like to thank all the subjects who have participated in this study.

Conflicts of Interest: The authors declare no conflicts of interest.

References

1. Thibos, L.N.; Applegate, R.A.; Schwiegerling, J.T.; Webb, R. Standards for Reporting the Optical Aberrations of Eyes. *J. Refract. Surg.* **2002**, *18*, 652–660.
2. López-Gil, N.; Howland, H.C. Measurement of the Eye's Near Infrared Wave-Front Aberration Using the Objective Crossed-Cylinder Aberroscope Technique. *Vision Res.* **1999**, *39*, 2031–2037. [CrossRef]
3. Walsh, G.; Howland, H.C.; Charman, W.N. Objective Technique for the Determination of Monochromatic Aberrations of the Human Eye. *J. Opt. Soc. Am. A* **1984**, *1*, 987–992. [CrossRef] [PubMed]
4. Marcos, S.; Diaz-Santana, L.; Llorente, L.; Dainty, C. Ocular Aberrations with Ray Tracing and Shack–Hartmann Wave-Front Sensors: Does Polarization Play a Role? *J. Opt. Soc. Am. A* **2002**, *19*, 1063–1072. [CrossRef]
5. Artal, P. Understanding Aberrations by Using Double-pass Techniques. *J. Refract. Surg.* **2000**, *16*, 560–562.
6. Liang, J.; Grimm, B.; Goelz, S.; Bille, J.F. Objective Measurement of Wave Aberrations of the Human Eye with the Use of a Hartmann–Shack Wave-Front Sensor. *J. Opt. Soc. Am. A* **1994**, *11*, 1949–1957. [CrossRef]
7. Hongbin, Y.; Guangya, Z.; Siong, C.F.; Feiwen, L.; Shouhua, W.A. Tunable Shack–Hartmann Wavefront Sensor Based on a Liquid-Filled Microlens Array. *J. Micromech. Microeng.* **2008**, *18*, 105017. [CrossRef]
8. Seifert, L.; Liesener, J.; Tiziani, H.J. The Adaptive Shack–Hartmann Sensor. *Opt. Commun.* **2003**, *216*, 313–319. [CrossRef]
9. Akondi, V.; Falldorf, C.; Marcos, S.; Vohnsen, B. Phase Unwrapping with a Virtual Hartmann–Shack Wavefront Sensor. *Opt. Express* **2015**, *23*, 25425–25439. [CrossRef]
10. Maeda, N. Wavefront Technology in Ophthalmology. *Curr. Opin. Ophthalmol.* **2001**, *12*, 294–299. [CrossRef]
11. Gobbe, M.; Guillon, M. Corneal Wavefront Aberration Measurements to Detect Keratoconus Patients. *Cont. Lens Anterior Eye* **2005**, *28*, 57–66. [CrossRef] [PubMed]
12. Dolgin, E. The Myopia Boom. *Nature* **2015**, *519*, 276–278. [CrossRef] [PubMed]

13. Vohnsen, B.; Carmichael, M.A.; Qaysi, S.; Sharmin, N. Hartmann–Shack Wavefront Sensing without a Lenslet Array Using a Digital Micromirror Device. *Appl. Opt.* **2018**, *57*, E199–E204. [CrossRef] [PubMed]
14. Lochocki, B.; Gambín, A.; Manzanera, S.; Irles, E.; Tajahuerce, E.; Lancis, J.; Artal, P. Single Pixel Camera Ophthalmoscope. *Optica* **2016**, *3*, 1056–1059. [CrossRef]
15. Carmichael, M.A.; Vohnsen, B. Analysing the Impact of Myopia on the Stiles-Crawford Effect of the First Kind Using a Digital Micromirror Device. *Ophthal. Physl. Opt.* **2018**, *38*, 273–280. [CrossRef] [PubMed]
16. Navarro, R.; Moreno-Barriuso, E. Laser Ray-Tracing Method for Optical Testing. *Opt. Lett.* **1999**, *24*, 951. [CrossRef] [PubMed]
17. Akondi, V.; Vohnsen, B. Myopic Aberrations: Impact of Centroiding Noise in Hartmann Shack Wavefront Sensing. *Ophthal. Physl. Opt.* **2013**, *33*, 434–443. [CrossRef] [PubMed]
18. Thibos, L.N.; Hong, X.; Bradley, A.; Cheng, X. Statistical Variation of Aberration Structure and Image Quality in a Normal Population of Healthy Eyes. *J. Opt. Soc. Am. A* **2002**, *19*, 2329–2348. [CrossRef]
19. Ryle, J.P.; Vohnsen, B.; Sheridan, J.T. Simultaneous Drift, Microsaccades, and Ocular Microtremor Measurement from a Single Noncontact Far-Field Optical Sensor. *J. Biomed. Opt.* **2015**, *20*, 027004. [CrossRef]
20. Rayner, K. Eye Movements in Reading and Information Processing: 20 Years of Research. *Psychol. Bull.* **1998**, *124*, 372–422. [CrossRef]
21. Otero-Millan, J.; Troncoso, X.G.; Macknik, S.L.; Serrano-Pedraza, I.; Martinez-Conde, S. Saccades and Microsaccades during Visual Fixation, Exploration, and Search: Foundations for a Common Saccadic Generator. *J. Vis.* **2008**, *8*, 21. [CrossRef] [PubMed]
22. Wong, Y.L.; Saw, S.M. Epidemiology of Pathologic Myopia in Asia and Worldwide. *Asia Pac. J. Ophthalmol.* **2016**, *5*, 394–402. [CrossRef] [PubMed]
23. Hiraoka, T.; Yamamoto, T.; Okamoto, F.; Oshika, T. Time Course of Changes in Ocular Wavefront Aberration after Administration of Eye Ointment. *Eye* **2012**, *26*, 1310–1317. [CrossRef] [PubMed]
24. Cox, M.A.; Toninelli, E.; Cheng, L.; Padgett, M.; Forbes, A. A High-Speed, Wavelength Invariant, Single-Pixel Wavefront Sensor with a Digital Micromirror Device. *Opt. Express* **2019**, in press.

MDPI

Article

Tunable Fiber Laser with High Tuning Resolution in C-band Based on Echelle Grating and DMD Chip

Jinliang Li [1], Xiao Chen [1,*], Dezheng Dai [1], Yunshu Gao [2], Min Lv [1] and Genxiang Chen [1,*]

1 College of Science, MINZU University of China, Beijing 100081, China; jinliangft@163.com (J.L.); daidezheng@126.com (D.D.); lvmin62589149@163.com (M.L.)
2 School of Electronic and Information Engineering, Beijing Jiaotong University, Beijing 100044, China; gaoyunshu@126.com
* Correspondence: xchen4399@126.com (X.C.); gxchen_bjtu@163.com (G.C.); Tel.: +86-01-68933910-523 (X.C.)

Received: 27 November 2018; Accepted: 4 January 2019; Published: 8 January 2019

Abstract: The tunable fiber laser with high tuning resolution in the C-band is proposed and demonstrated based on a digital micromirror device (DMD) chip and an echelle grating. The laser employs a DMD as a programmable wavelength filter and an echelle grating with high-resolution features to design a cross-dispersion optical path to achieve high-precision tuning. Experimental results show that wavelength channels with 3 dB-linewidth less than 0.02 nm can be tuned flexibly in the C-band and the wavelength tuning resolution is as small as 0.036 nm. The output power fluctuation is better than 0.07 dB, and the wavelength shift is below 0.013 nm in 1 h at room temperature.

Keywords: tunable fiber laser; echelle grating; DMD chip

1. Introduction

Tunable lasers as a powerful tool have been widely applied in spectroscopy, photochemistry, biomedicine, and optical communications for decades. For example, in dense wavelength division multiplexing (DWDM) optical communication, tunable lasers can not only replace multiple fixed-wavelength lasers to save the operation cost but also realize the remote dynamic allocation of networks resources. The number of wavelength channels in C-band determines the information transmission capacity in networks. Therefore, how to improve narrow-linewidth channels with a high tuning accuracy from laser sources has been receiving an increasing amount of attention from researchers and network service vendors.

To date, various technologies have been proposed and implemented to realize tunable filters in laser sources, including fiber Bragg grating (FBG), Fabry–Perot (F–P) cavity, acousto-optics, interferometer, liquid crystal on silicon (LCoS), etc. FBG can be tuned easily through either heating or applying strain along the device. For example, it is reported that FBG can achieve 0.2 nm/V tuning accuracy from 1555–1565 nm driven by direct current (DC) voltage of multilayer piezoelectric transducers [1]. However, FBG-based tunable lasers are affected by the environment fluctuation, resulting in a high packaging cost and limited tuning range. The fiber-optic self-seeding F–P cavity achieves a wide range of single longitudinal modes tuning from 1153.75 to 1560.95 nm with a tuning step of 1.38 nm [2]. Avanaki et al. investigate a fiber Fabry–Pérot tunable filter using a well-established optimization method, simulated annealing (SA), to achieve maximum amplitude for the Fourier transformed peaks of the photodetected interferometric signal [3]. Furthermore, Y. Ding implemented a small-scale tuning with the accuracy of approximately 0.6 nm by using micro-ring Mach–Zehnder interferometers [4]. These technologies generally need additional matching devices, like an F–P laser, saturable absorber-based filters, which makes them complex and expensive to commercialize. Nowadays, a LCoS spatial light modulator as a programmable filter produced by Very-Large-Scale-Integrated (VLSI) technology has been applied to laser systems [5]. A digital micromirror device (DMD), another Opto-VLSI processor has also been attempted in a non-projection field. In 2006, Chuang and Lo proposed a spectral synthesis method with a spectral tuning accuracy of 0.076 nm/pixel in the C-band based on a DMD chip [6]. In 2009, W. Shin used the DMD-based tunable laser system as light sources for the optical time domain reflectometry, with a tuning range of 1525–1562 nm, and an improved laser tuning accuracy of 0.1 nm [7]. Our research group also reported a multi-wavelength tunable fiber laser based on a DMD chip with a step of 0.055 nm [8].

Echelle gratings are a special type of blazed gratings featured by a large blazing angle of grooves and often operate at high diffraction orders to obtain high dispersion. They are different to conventional gratings [9–12]. An echelle grating splits the radiant energy into a multitude of diffraction orders that overlap in the narrow interval of the grating diffraction angle. Therefore, in practical application, an additional order separator like the prism or grating, whose dispersion direction is perpendicular to that of an echelle grating are inserted to separate the overlapping orders. By focusing the two-fold dispersed radiation, a two-dimensional spectrum is produced, thus achieving an applicable high-resolution spectrum. So far, echelle gratings are mainly applied in ultraviolet and visible high-resolution spectrometers [10,11].

In this work, we first apply an echelle grating into a DMD-based tunable laser to realize the high tuning resolution in C-band. The echelle-based tunable fiber laser is designed for a cross-dispersion structure of a closed-loop fiber system. The laser wavelength was tuned in the range of 1540–1560 nm with a tuning step of 36 pm. The 3dB-linewidth of the signals was less than 0.02 nm. The side mode suppression ratio (SMSR) reaches 40 dB, and the maximum output power was 7.5 dBm.

2. Echelle Grating and System Design

The spectral order of an echelle grating is the result of mutual modulation of multi-slit interference and single-slit diffraction. The echelle equation is expressed as:

$$m\lambda = d(\sin\alpha + \sin\beta)\cos\gamma \tag{1}$$

where m, λ, and d are the diffraction order, wavelength, and grating constant, respectively. α, β and γ are the incident angle, corresponding diffraction angle, and off-axis angle. As shown in Figure 1a, θ_B is the blaze angle of an echelle grating and θ is the incident angle to the facet. So, the relation of angles is written as:

$$\alpha = \theta_B + \theta, \beta = \theta_B - \theta \tag{2}$$

Substituting Equation (2) into (1), the diffraction of an echelle grating is characterized as follows:

$$m\lambda = 2d\sin\theta_B\cos\theta\cos\gamma \tag{3}$$

An echelle grating has the maximum diffraction efficiency only when the Littrow condition is satisfied, that is the incidence is at the blaze angle. On both sides of the blaze angle, the diffraction efficiency of a grating decreases rapidly as θ increases. However, the strict Littrow condition leads to the difficulty in the arrangement of the actual optical path. Therefore, a quasi-Littrow structure is usually employed with the incident ray at an off-axis angle γ from the principal section of a grating, as shown in Figure 1b. The condition of the quasi-Littrow configuration is:

$$\theta = 0, \gamma \neq 0 \tag{4}$$

Therefore, the echelle grating equation under the quasi-Littrow condition is:

$$m\lambda = 2d\sin\theta_B\cos\gamma \tag{5}$$

The free spectral range $\triangle\lambda_{\text{SFR}}$:

$$\triangle \lambda_{\text{SFR}} = \frac{\lambda}{m} = \frac{\lambda^2}{2d \sin\theta_B \cos\gamma} \tag{6}$$

The range of the dispersion angle of m-order:

$$\triangle \theta = \triangle\lambda_{SFR} = \frac{\mathrm{d}\theta}{\mathrm{d}\lambda} = \frac{2\tan\theta_B}{m} \tag{7}$$

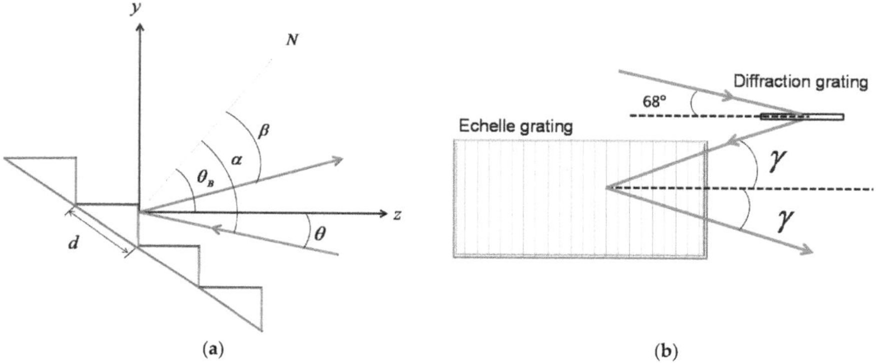

Figure 1. (**a**) Echelle grating working characteristics, (**b**) off-axis angles γ (the incident angle of diffraction grating is 68°).

It can be seen from Equations (5)–(7) that an echelle grating has the following features: (1) The free spectral range is small and the spectral order is seriously overlapped. Therefore, it is necessary to use auxiliary dispersion elements for cross-dispersion to obtain a two-dimensional spectrum. (2) The angular dispersion is so high that the wavelength resolution is greatly improved. (3) The dispersion angle of one order is small, and the wavelengths in the free spectral range of each stage are concentrated near the blazed order, so an echelle grating can blaze in the entire band.

In Figure 2, the two-dimensional cross-dispersion is realized by a diffraction grating and an echelle grating with main sections that are perpendicular to each other. As we know, echelle gratings are, to date, mainly applied in ultraviolet (UV) and visible (VIS) spectrometers, so most of the prisms are used as auxiliary dispersers placed before or after an echelle grating to achieve cross-dispersion. In our work, the laser operates in the C-band and the prism glass material shows strong absorption in infrared. Therefore, a diffraction grating is adopted to replace the prisms in fiber lasers.

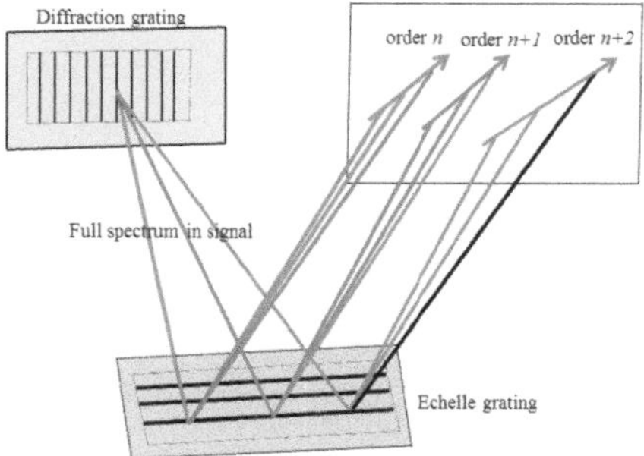

Figure 2. Schematic of cross-dispersion of the diffraction grating and the echelle grating.

Figure 3 demonstrates the tunable laser structure employing a DMD chip as a programmable filter in bulk optics and a fiber resonator with an erbium-doped fiber amplifier (EDFA). The lasing process in a fiber cavity is achieved by optical pumping and erbium gain. The bulk optics obtain the high-precision mode selection by an echelle grating and a 0.55″ DMD in the experiment. The detailed working principle of a 0.55″ DMD and its diffraction efficiency have been analyzed in [13]. The EDFA emits the amplified spontaneous emission spectrum (ASE) signals from 1530–1560 nm. After a 90/10 optical fiber coupler, 90% ASE light energy returns into a ring and then continues to be coupled into the bulk optics via a circulator and an optical fiber collimator. The bulk optics consists of two cylindrical lenses, a diffraction grating, an echelle grating, and a DMD chip. The fiber collimator and the 1200 line/mm diffraction grating are located at the front and the rear focal planes of lens (f_0 = 100 mm), respectively. The diffraction grating and the 79 line/mm echelle grating are separated by 100 mm. In order to ensure that the echelle grating adheres to the quasi-Littrow condition, the incident beam is arranged at an off-axis angle γ so that the diffracted beam and the incident beam are in the same horizontal plane. The cylindrical lens 1 (f_1 = 150 mm) and cylindrical lens 2 (f_2 = 100 mm) are 50 mm and 100 mm from the echelle grating, respectively. Therefore, the busbars of two cylindrical lenses are perpendicular to each other, and the two dispersion directions after two gratings are collimated, respectively. The DMD is at the back focal plane of two cylindrical lenses, as shown in Figure 4. By uploading steering holograms onto the DMD controlled by remote software, any waveband of ASE spectra can be routed and coupled into the optical system along the original path, and the others are dropped out with dramatic attenuation, thereby achieving the laser longitudinal mode selection and wavelength tuning. The selected wavebands through the collimator and circulator returning into a ring cavity are amplified by EDFA, leading, after several recirculations, to high-quality single-mode laser generation.

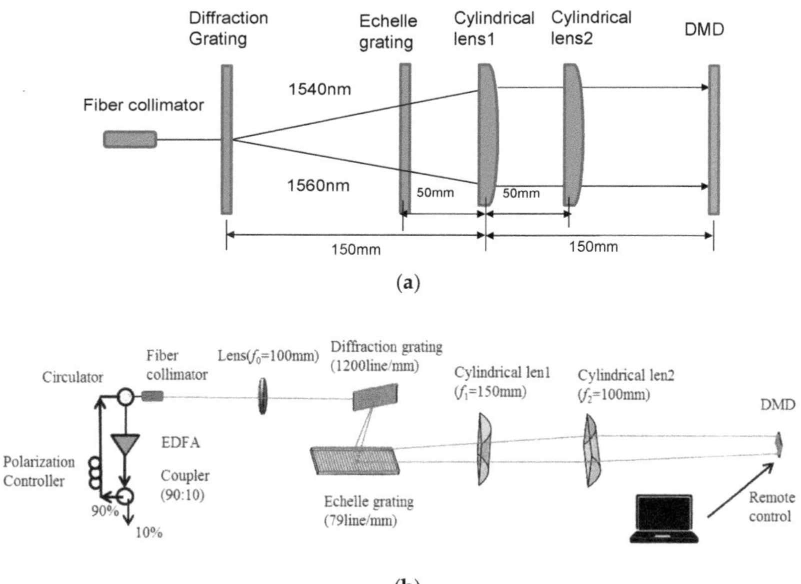

Figure 3. (**a**) Diagram of a cross-dispersion tunable laser system. (**b**) Layout of wavelength selective path.

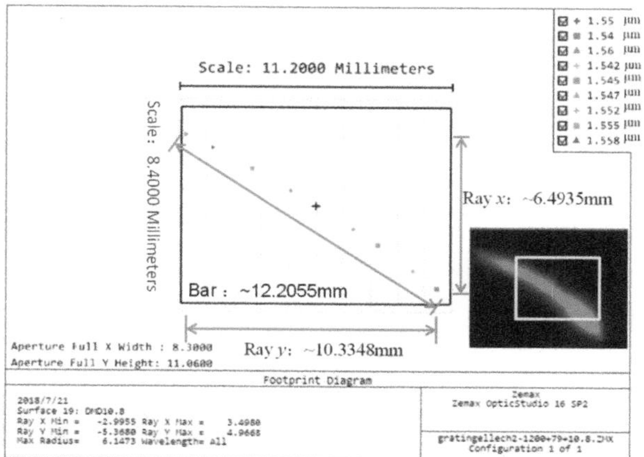

Figure 4. Distribution of the dispersion bar on the digital micromirror device (DMD) surface simulated by Zemax OpticStudio. The inset is the experimental pattern (rectangular box is illustrated as a DMD).

The off-axis arrangement greatly influences the laser tuning range and accuracy. We optimize the off-axis angle γ of the echelle grating (79 line/mm) for the laser system. According to Equation (5), m = 15 and λ = 1550 nm are selected as the calibration blazed wavelength, and the corresponding γ under the quasi-Littrow condition is calculated as 18.05°. Using Zemax OpticStudio software, we design the optical system to analyze the beam distribution on the DMD surface. The simulation results illustrate the length of the two-dimensional dispersion strip is 12.2 mm in Figure 4, matching with the experimental pattern in the inset of Figure 4. The 0.55″ DMD receiving wavelength range is around 20 nm from 1540–1560 nm and is limited by the DMD size. The tuning accuracy of the laser wavelength is 0.0177 nm/pixel, in theory. Considering the used echelle grating has a wide working range from UV to 25 µm, this laser system is convenient to be extended in the 2 µm-band, which has potential applications in the biomedical domain [14,15].

3. Experimental Results

When the optical loop is closed, Figure 5 shows a typical laser signal with the center of the wavelength at 1546.733 nm when the pump power is 120 mW. The power of the laser output is around 7.5 dBm, the 3 dB-linewidth is less than 0.02 nm (limited by the resolution of the YOKOGAWA spectrum analyzer, Yokogawa Test & Measurement Corporation, Tokyo, Japan), and the SMSR exceeds 40 dB.

Different holograms are loaded onto the DMD chip, each hologram corresponds to a different selected wavelength. Each selected wavelength is amplified by EDFA to achieve lasing. Figure 6 is the measured outputs of the echelle-grating-based fiber laser tuning from 1542 to 1558 nm by remotely uploading the 8 × 768 pixel-holograms at different positions along the DMD active window when the threshold pumping power is 28 mW. It demonstrates an excellent tuning capability. Notice that the range of the actual tuning wavelength is a little wider than 16 nm. The wavelength outside the tuning range requires a higher threshold power to lasing due to the off-axis angle and the influence of stray light.

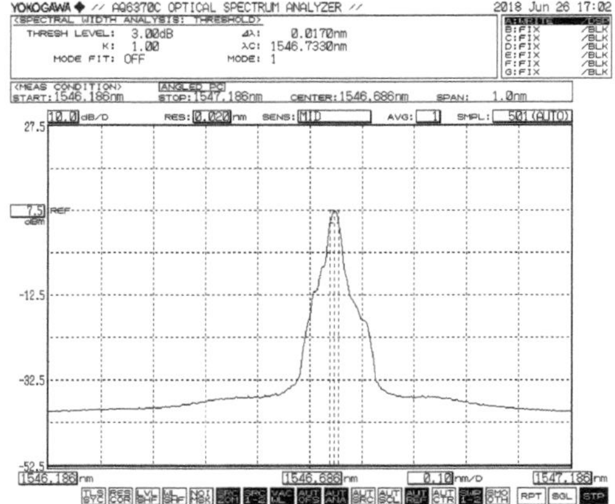

Figure 5. Typical laser signal with the wavelength centered at 1546.733 nm.

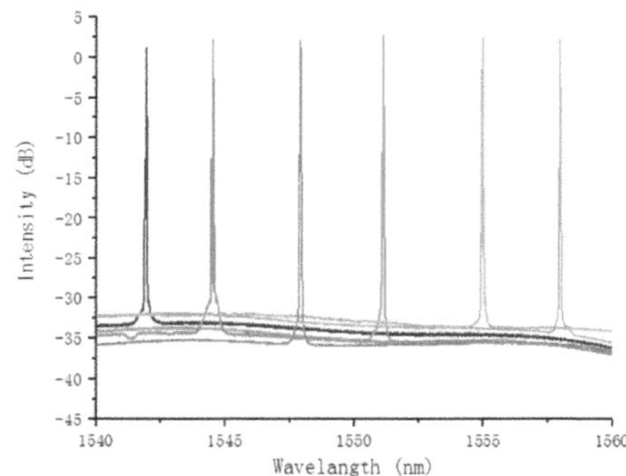

Figure 6. Coarse tuning characteristics of fiber laser in the range 1542–1558 nm.

Figure 7 is the fine tuning characteristics of laser outputs with the fine tuning accuracy 0.036 nm. We modulate the selected wavelength each time by moving 2-pixels on the hologram. The tuning accuracy corresponding to each pixel is related to the number of DMD pixels covered by the ASE spectrum on the surface of the DMD. Note that the tuning accuracy can be further improved by employing a DMD with a smaller pixel size, like the DLP2010NIR (Texas Instruments Incorporated, Dallas, TX, USA. Each pixel size is 5.4 μm). The shoulders on both sides of the laser spectrum may be due to self-phase modulation or other nonlinear phenomena arising from a high-level of output power [16].

Figure 8 shows the drift of wavelength (dotted line) and the fluctuation of peak power (solid line) at the pump power 40 mW during 1-h observation at the center wavelength of 1546 nm. The maximum wavelength drift is less than 0.013 nm and the maximum peak power fluctuation is 0.07 dB at room temperature. The linewidth is better than that reported in [5] (0.05 nm) and [8] (0.02 nm), and the maximum peak power fluctuation is better than that in [8] (0.25 dB). Compared with other tunable lasers with the same tuning mechanism, the laser output stability has been further improved.

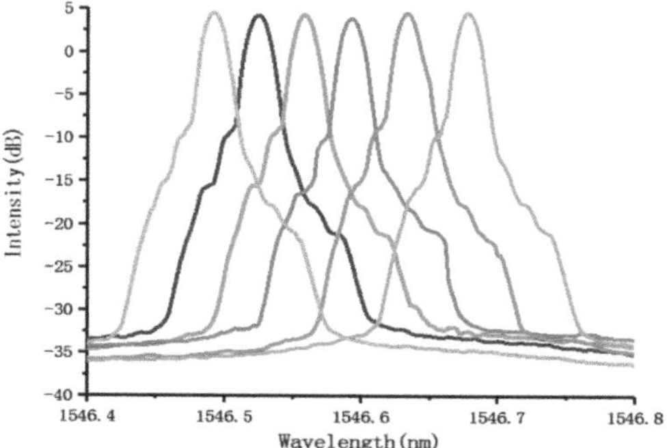

Figure 7. Fine tuning characteristics of the laser system from 1546.4–1546.8 nm with a tuning step of 0.036 nm.

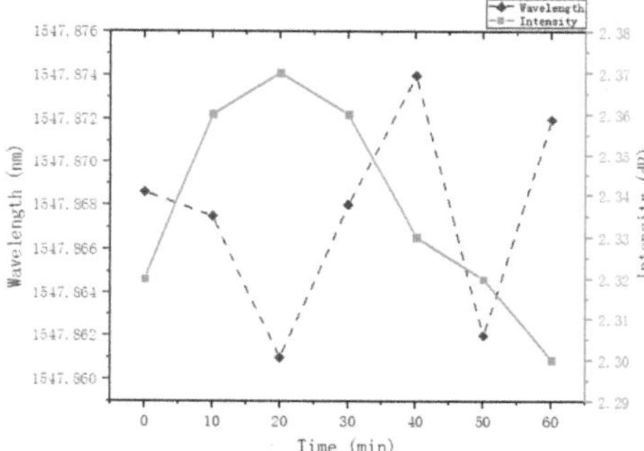

Figure 8. The shift of the center wavelength (dash line) and the fluctuation of laser powers (solid line) within 1 h.

Finally, due to the off-axis angles introduced into the tunable fiber laser, the aberration-like coma and astigmatism influences the tuning range and accuracy. Therefore, we will continue to optimize the optical path and reduce the stray light effect caused by an echelle grating in the follow-up work, which will be helpful to further improve the tuning property of devices. Also, loading the modulation algorithm on the DMD is an attractive solution, and our research process in the future will also consider using algorithms to further improve the performance of tunable fiber lasers.

4. Conclusions

The C-band tunable fiber laser based on a DMD chip and an echelle grating is proposed and demonstrated experimentally. The laser employs a DMD as a programmable wavelength filter and an echelle grating with high-resolution features to design a cross-dispersion optical path to achieve high-precision tuning. The optimal off-axis angle of an echelle grating under the quasi-Littrow condition is simulated and analyzed in detail. Experimental results show that wavelength channels are tuned in the range of 1542–1558 nm with a tuning step of 0.036 nm. The 3 dB-linewidth of the signals is less than 0.02 nm, the SMSR reaches 40 dB, and the maximum output power is 7.5 dBm. At room temperature, the output power fluctuation is better than 0.07 dB in 1 h, and the wavelength shift is below 0.013 nm.

Author Contributions: X.C. and G.X. conceived and designed the experiments; J.L. performed the experiments; J.L., Y.G. and D.D. analyzed the data; G.X. and M.L. contributed reagents/materials/analysis tools; J.L. and X.C. wrote the paper.

Acknowledgments: We acknowledge financial support from the National Science Foundation of China (Grant Nos: 61675238, 61275052), National Key Scientific Instrument and Equipment Development Project (Grant No: 61627814).

Conflicts of Interest: The authors declare no conflict of interest.

References

1. Inui, T.; Komukai, T.; Nakazawa, M. Highly efficient tunable fiber Bragg grating filters using multilayer piezoelectric transducers. *Opt. Commun.* **2001**, *190*, 1–4. [CrossRef]
2. Yeh, C.H.; Shih, F.Y.; Wang, C.H.; Chow, C.W.; Chi, S. Cost-effective wavelength-tunable fiber laser using self-seeding Fabry-Perot laser diode. *Opt. Express* **2008**, *16*, 435–439. [CrossRef] [PubMed]
3. Avanaki, M.R.N.; Bradu, A.; Trifanov, I.; Ribeiro, A.B.L.; Hoijatoleslami, A.; Podoleanu, A.G. Algorithm for Excitation Optimization of Fabry–Pérot Filters Used in Swept Sources. *IEEE Photon. Tech. Lett.* **2013**, *25*, 472–475. [CrossRef]
4. Ding, Y.; Pu, M.; Liu, L.; Xu, J.; Peucheret, C.; Zhang, X.; Huang, D.; Ou, H. Bandwidth and wavelength-tunable optical bandpass filter based on silicon microring-MZI structure. *Opt. Express* **2011**, *19*, 6462–6470. [CrossRef] [PubMed]
5. Xiao, F.; Alameh, K.; Lee, T.T. Opto-VLSI-based tunable single-mode fiber laser. *Opt. Express* **2009**, *17*, 18676–18680. [CrossRef] [PubMed]
6. Chuang, C.H.; Lo, Y.L. Digital programmable light spectrum synthesis system using a digital micromirror device. *Appl. Opt.* **2006**, *45*, 8308–8314. [CrossRef] [PubMed]
7. Shin, W.J.; Yu, B.A.; Lee, Y.L.; Noh, Y.C.; Ko, D.K.; Oh, K. Wavelength tunable optical time-domain reflectometry based on wavelength swept fiber laser employing two-dimensional digital micro-mirror array. *Opt. Commun.* **2009**, *282*, 1191–1195. [CrossRef]
8. Ai, Q.; Chen, X.; Tian, M.; Yan, B.B.; Zhang, Y.; Song, F.J.; Chen, G.X.; Sang, X.Z.; Wang, Y.Q.; Xiao, F.; Alameh, K. Demonstration of multi-wavelength tunable fiber lasers based on a digital micromirror device processor. *Appl. Opt.* **2015**, *54*, 603–607. [CrossRef] [PubMed]
9. Bykov, S.V.; Sharma, B. High-Throughput, High-Resolution Echelle Deep-UV Raman Spectrometer. *Appl. Spectrosc.* **2013**, *67*, 873. [CrossRef] [PubMed]
10. Zhang, R.; Bayanheshig, L.Y.; Yin, L.; Li, X.; Cui, J.; Yang, J.; Sun, C. Wavelength calibration model for prism-type echelle spectrometer by reversely solving prism's refractive index in real time. *Appl. Opt.* **2016**, *55*, 4153–4158. [CrossRef]
11. Luo, D.; Taphanel, M.; Längle, T.; Beyerer, J. Programmable light source based on an echellogram of a supercontinuum laser. *Appl. Opt.* **2017**, *56*, 2359–2367. [CrossRef] [PubMed]
12. Yin, L.; Bayanheshig, J.Y.; Lu, Y.; Zhang, R.; Sun, C.; Cui, J. High-accuracy spectral reduction algorithm for the échelle spectrometer. *Appl. Opt.* **2016**, *55*, 3574–3581. [CrossRef] [PubMed]
13. Chen, X.; Yan, B.B.; Song, F.J.; Wang, Y.Q.; Xiao, F.; Alameh, K. Diffraction of digital micromirror device gratings and its effect on properties of tunable fiber lasers. *Appl. Opt.* **2012**, *51*, 7214–7220. [CrossRef] [PubMed]
14. Shin, W.; Lee, Y.L.; Yu, B.A.; Noh, Y.C.; Ahn, T.J. Wavelength-tunable thulium-doped single mode fiber laser based on the digitally programmable micro-mirror array. *Opt. Fiber Technol.* **2013**, *19*, 304–308. [CrossRef]
15. Billaud, A.; Shardlow, P.C.; Clarkson, W.A. Wavelength-flexible thulium-doped fiber laser employing a digital micro-mirror device tuning element. In Proceedings of the Conference on Lasers and Electro-Optics, OSA Technical Digest, San Jose, CA, USA, 5–10 June 2016.
16. Prabhu, M.; Kim, N.S.; Ueda, K. Simultaneous Double-Color Continuous Wave Raman Fiber Laser at 1239 nm and 1484 nm Using Phosphosilicate Fiber. *Opt. Rev.* **2000**, *7*, 277–280. [CrossRef]

 *micromachines*

Article

Multiple Laser Stripe Scanning Profilometry Based on Microelectromechanical Systems Scanning Mirror Projection

Gailing Hu [1,2], Xiang Zhou [1,2,3,*], Guanliang Zhang [1,2], Chunwei Zhang [1,2], Dong Li [1,2] and Gangfeng Wang [4]

1. State Key Laboratory for Manufacturing Systems Engineering, Xi'an Jiaotong University, Xi'an 710049, China; hugl@mail.xjtu.edu.cn (G.H.); zgl862203570@stu.xjtu.edu.cn (G.Z.); zcw198811@163.com (C.Z.); ohyailidong@126.com (D.L.)
2. School of Mechanical Engineering, Xi'an Jiaotong University, Xi'an 710049, China
3. School of Food Equipment Engineering and Science, Xi'an Jiaotong University, Xi'an 710049, China
4. Key Laboratory of Road Construction Technology and Equipment of MOE, Chang'an University, Xi'an 710064, China; wanggf@chd.edu.cn

* Correspondence: zhouxiang@mail.xjtu.edu.cn

Received: 10 December 2018; Accepted: 10 January 2019; Published: 16 January 2019

Abstract: In traditional laser-based 3D measurement technology, the width of the laser stripe is uncontrollable and uneven. In addition, speckle noise in the image and the noise caused by mechanical movement may reduce the accuracy of the scanning results. This work proposes a new multiple laser stripe scanning profilometry (MLSSP) based on microelectromechanical systems (MEMS) scanning mirror which can project high quality movable laser stripe. It can implement full-field scanning in a short time and does not need to move the measured object or camera. Compared with the traditional laser stripe, the brightness, width and position of the new multiple laser stripes projected by MEMS scanning mirror can be controlled by programming. In addition, the new laser strip can generate high-quality images and the noise caused by mechanical movement is completely eliminated. The experimental results show that the speckle noise is less and the light intensity distribution is more even. Furthermore, the number of pictures needed to be captured is significantly reduced to $1/N$ (N is the number of multiple laser stripes projected by MEMS scanning mirror) and the measurement efficiency is increased by N times, improving the efficiency and accuracy of 3D measurement.

Keywords: 3D measurement; laser stripe width; vibration noise; MLSSP; MEMS scanning mirror

---

1. Introduction

3D scanning using a structured light projection is widely applied to the measurement of geometric parameters and 3D reconstruction of object surfaces in many fields, including industrial inspection [1–3], biomedical treatments [4], culture heritage digitization [5] and food detection [6,7]. This method has several advantages, including noncontact measurement, large measurement ranges, high speed and high accuracy.

Structured light can be divided into two categories: coded-pattern and fixed-pattern light [8]. In coded-pattern light, the most is the Digital Light Processing (DLP) projection, which is widely used in optical measurement because DLP projection is a programmable pattern [9–11]. DLP usually projects sinusoidal fringes to obtain the information modulated by the surface of the object, thereby the three-dimensional reconstruction is implemented. This method usually takes a few seconds to implement full-field scanning [12–14]. However, the DLP based method is sensitive to measurement environment and the surface reflectivity of the object. The reliability of DLP projection is reduced in the case where the surface reflectivity difference of the measured object is large or the measurement environment is complicated [15].

Laser is mainly taken as the light source of a fixed-pattern light [8]. In this laser-based 3D reconstruction system, a laser stripe is projected onto an object and then a camera acquires a series of laser stripe images while the object or the laser stripe is moving forward. The laser stripe is modulated by the shape of the object. Thus, using optical triangulation is possible to calculate 3D information of the object. Due to the intensity and information concentration of the laser stripes, even if the surrounding light is not controlled or the surface of the object to be measured is complex, the measurement results of this method are rarely affected by the environment and the measured object. Therefore, laser-based 3D reconstruction methods can be applied to various industrial environments. Since this method mainly relies on the extraction of the center of the laser strip to obtain the surface information of the measured object, better quality laser stripe can be used to achieve higher measurement accuracy. However, the width of the traditional laser stripe is difficult to reduce and the light intensity distribution is not even, affecting the lateral resolution [16,17]. In addition, the measuring time of this method is mainly subject to the scanning mechanism and the vibration caused by the mechanical movement during the scanning also affects the measurement accuracy [18]. These shortcomings greatly reduce the efficiency and accuracy of measurement results.

In view of the above, it is significant to improve the accuracy of 3D measurement by replacing the traditional line laser scanning with a projector which can project high quality laser stripes. The laser micro-mirror scanning appeared about 40 years ago based on the fact that laser is taken as a light source and micro-mirror as a light modulator [19]. This technology is implemented by microelectromechanical systems (MEMS) manufacturing process which can realize the scanning of laser beam through different control strategies to form a two-dimensional projection image [20,21]. With the development of MEMS, laser micro-mirror scanning technology has found wide applications in engineering in recent years [22–25]. MEMS scanning mirror has been used in optical coherence tomography (OCT) scans [26,27], time of flight (ToF) cameras [28], 3D confocal scanning microscopes [29–31] and other fields of measurement [32–34], however, there are few reports of its use in 3D measurement. In this work, a novel 3D measurement method, called multiple laser stripe scanning profilometry (MLSSP) based on MEMS scanning mirror projection, is proposed, which can solve the problems presented above. Compared with conventional laser scanning method, the proposed method has many advantages. The brightness, width, period and position of the new multiple laser stripes projected by MEMS scanning mirror can be controlled by programming, which can generate high-quality measurement images, thereby the MLSSP based on MEMS scanning mirror can improve the accuracy of 3D measurement. The MLSSP is capable of completing full-field scanning measurement in a short time without moving the projector or object, completely eliminates the measurement error caused by the vibration and the measurement efficiency is improved by N times in contrast to the traditional laser-based 3D measurement method. In addition, the proposed method is less affected by industrial environment and surface reflectivity. Due to the robustness, high efficiency and accuracy, this method can be applied to measure objects in various industrial measurement environments including obviously changed surface reflectivity of the measured object and other complicated environment.

This paper is organized as follows: In Section 2, the methods and principles of MLSSP are introduced. In Section 3, experiments on laser stripe performance and 3D measurement are conducted, followed by the discussion of the results and conclusions will be briefly described in Section 4.

2. Methods and Principles

In order to improve the scanning speed and reduce the number of pictures in 3D measurement, the MEMS scanning mirror is used to project simultaneously multiple parallel laser stripes. If the number of laser stripes projected is N, the number of images to be captured will be reduced to $1/N$ and the scan time will also be reduced to $1/N$ compared to the single laser stripe scanning. The scanning using MLSSP is generally performed in a direction perpendicular to the laser stripe at a fixed interval. By taking the number of laser stripes $N = 4$ as an example, multiple laser stripes scanning process is shown in Figure 1. Controlled by the multiple laser stripe coding method, MEMS scanning mirror projects four parallel laser stripes at time t_i,t_{i+1},t_{i+2} $\cdots\cdots$ with an equal time intervals. Its projection on the continuous surface is in the form of four unoverlapped curves at a certain interval. It only needs to scan the distance of Δx to implement full-field scanning instead of $4\Delta x$ by single laser stripe, wherein Δx is the distance of the adjacent laser stipe in the reference surface.

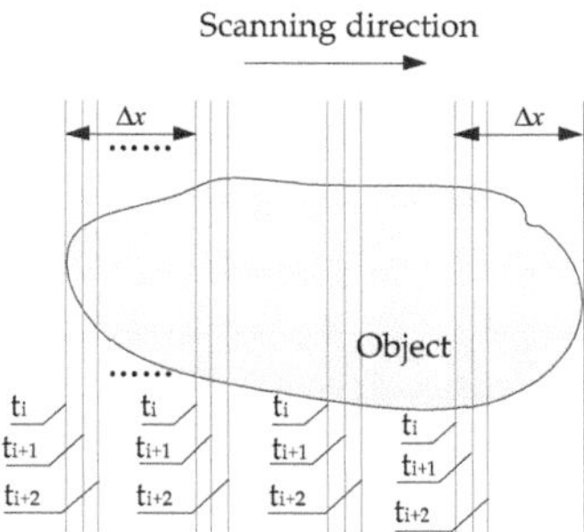

Figure 1. Scanning process of multiple laser stripe scanning profilometry (MLSSP) taking the number of laser stripes of $N = 4$.

Multiple laser stripe scanning profilometry involves two key techniques, namely multiple laser stripe coding and 3D reconstruction. This section introduces measurement methods and principles different from traditional laser-based 3D measurement technology.

2.1. Multiple Laser Stripe Coding Method

Figure 2 shows the driving signals generating multiple laser stripes projected by MEMS scanning mirror. When the laser incident onto micro-mirror, the micro-mirror fast axis performs simple harmonic motion under the control of a sinusoidal signal to achieve horizontal scanning. The micro-mirror slow axis implements the longitudinal scanning under the control of the sawtooth signal.

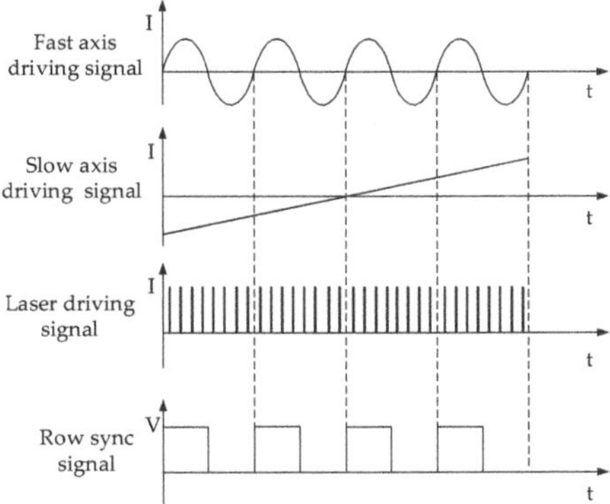

Figure 2. Driving signals generating multiple laser stripes projected by microelectrochemical systems (MEMS) scanning mirror.

The number, position and output light intensity of the laser stripes are controlled by laser driving pulse signals. There are four equally spaced laser impulse signals in the half cycle of the micro-mirror fast axis driving signal. After reflecting by micro-mirror, each row pixel points of spatial projection field are evenly spaced to generate four laser stripes with equal spacing.

Step synchronization signal controls the synergy between the laser and fast axis of micro-mirror. After the micro-mirror fast axis completes one cycle of scanning, the slow axis undergoes a slight angular deflection in the vertical direction and achieves accurate step-wise scanning under the control of the sawtooth wave signal, which in turn triggers the fast axis to scan in the next cycle. Figure 3 is one of the multiple laser stripe images projected by MEMS scanning mirror when setting the number of laser stripes N as 4.

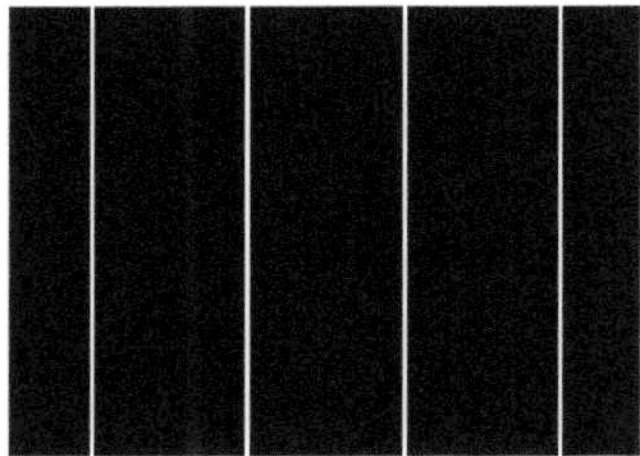

Figure 3. A four-laser stripe image projected by MEMS scanning mirror.

2.2. 3D Reconstruction Mechanism

MEMS scanning mirror projects multiple laser stripes onto the object and two cameras capture the deformed laser stripe images. In the experiment, the camera is MV-EM1200M produced by the Microvision of China, with a resolution of 1280 × 960 pixels, pixel size is 3.75 μm × 3.75 μm, equipped with 8 mm fixed lens. The projector is a MEMS scanning mirror driven by a signal control board with two-axis two-dimensional scanning electromagnetic micro-mirror. Figure 4a shows the internal optical path structure of the MEMS scanning mirror. The laser stripe captured by the camera is not a single pixel and hence the centerline of laser stripe needs to be extracted by centerline extraction method, then it is also required to carry out stereo matching and then 3D point cloud images of the object can be obtained. Figure 4b shows a representative schematic diagram of MLSSP. In order to avoid staggered superposition of adjacent laser stripes, the distance between adjacent laser stripes is correspondingly set wider when the depth variation of the measuring object is larger. Taking the left camera as an example, the most suitable distance between adjacent laser stripes is calculated and the number of projected multiple laser stripes will be determined.

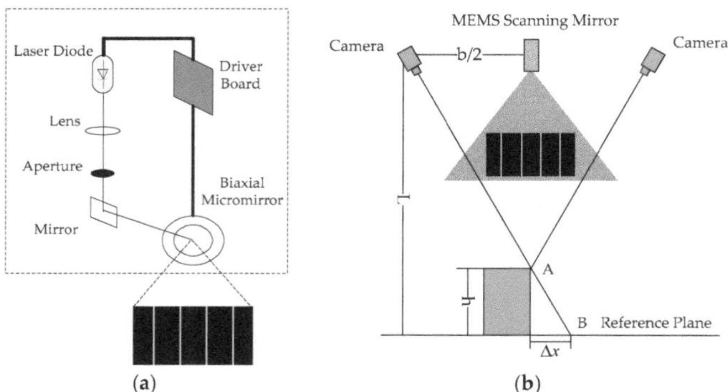

Figure 4. (**a**) Internal structure of the MEMS scanning mirror. (**b**) Schematic diagram of MLSSP.

The MEMS scanning mirror is placed at the center of the horizontal connection of the two cameras in the binocular system. It is known that the distance between the two cameras is b, the optical distance between the camera's optical center and the projection device is $b/2$ and the measurement distance is L. The physical size of the adjacent laser stripe distance is Δx and the maximum step height that can be measured is h. Points A and B are located on the adjacent two laser stripes in the reference plane. According to this triangular relationship, the physical size of the adjacent laser stripe distance is

$$\Delta x = \frac{bh}{2L} \tag{1}$$

The resolution of MEMS mirror projection is (u_p, v_p), the parameter of the projection ratio is R_P and the representation of the projection ratio is

$$R_P = \frac{L}{m} \tag{2}$$

where m is the maximum physical size of the long side of the projected image.

Figure 5 shows the geometry model of the proposed method, by taking the number of laser stripes $N = 4$ as an example. A laser source of MEMS mirror scans to generate multiple laser stripes A_1, A_2, A_3 and A_4 in the reference plane. One laser stripe is emitted from O, passes through a pixel E in the MEMS mirror plane and falls on a point A_1 in a reference plane. The adjacent laser stripe passes through a pixel point F in the MEMS mirror plane and falls at a point A_2 in a reference plane. The length of MN is m with the measurement distance of L, the pixel pitch size EF of the adjacent laser stripe in the MEMS projector is Δp that can be calculated by:

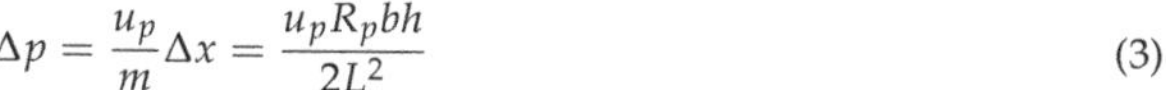

$$\Delta p = \frac{u_p}{m}\Delta x = \frac{u_p R_p bh}{2L^2} \tag{3}$$

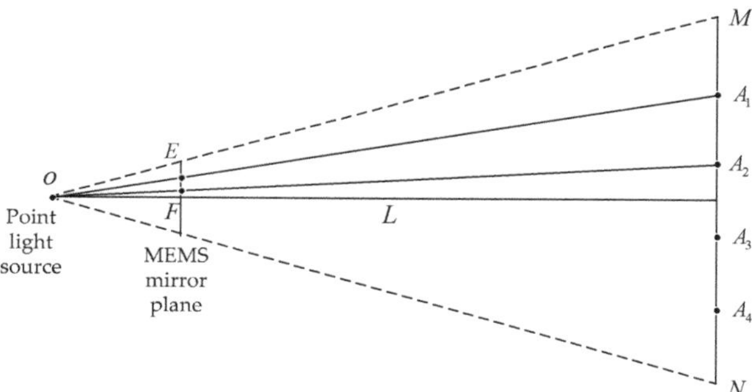

Figure 5. Geometry model of MLSSP.

In this way, if the MEMS scanning mirror projects N laser stripes and there are N laser stripe information in each line of the pictures captured by the left and right cameras, these laser strips in left and right images are matched in order. However, the laser stripe may be missing or broken when measuring the edge of the object, as shown in Figure 6. In this case, if the stereo matching is performed according to the principle of simple matching order, the matching of the laser stripes are difficult and results may be wrong. In order to solve this problem, the fixed area needs to be divided for each laser stripe.

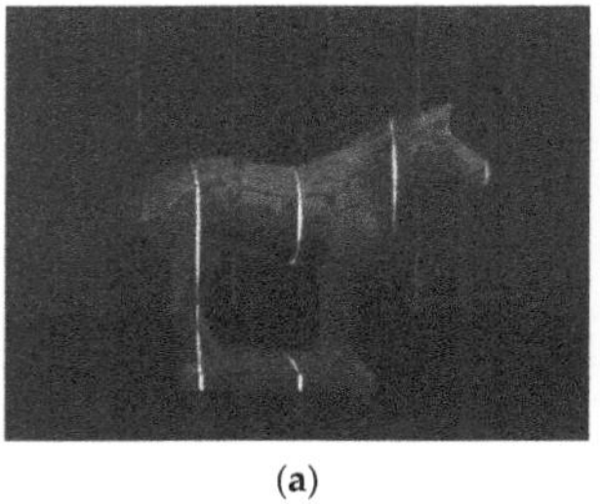
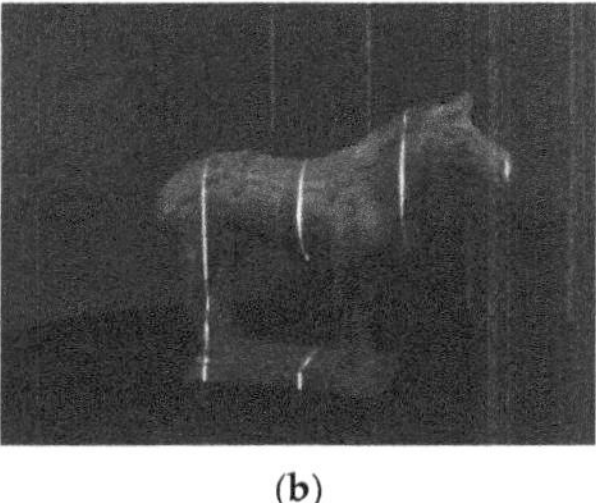

(a) (b)

Figure 6. Missing and damaged laser stripes: (**a**) Left image and (**b**) Right image.

When the surface depths of the measurement object are different, each laser stripe has a certain range in the image captured by camera. So the multiple laser stripe image can be divided into N areas. The ith laser stripe is always fixed in the ith area, then the left and right images can be matched according to the corresponding laser stripe area.

The following section describes the method of dividing laser stripes area by taking the number of laser stripes $N = 4$ as an example. Firstly, place the white flat plate parallel to two optical centers of the cameras with a measurement distance of L and then capture the laser stripe image by the camera. Calculate the longitudinal pixel coordinates of four laser stripe centerlines in the image are xn_1, xn_2, xn_3 and xn_4. Secondly, translate the white calibration plate to the position with the measurement distance of L–h, capture the laser stripe image, calculate the longitudinal pixel coordinates of the four laser stripe centerline in the distance of L–h as xf_1, xf_2, xf_3 and xf_4. Then the fixed area of each laser stripe in the image respectively are $(xn_1, xf_1), (xn_2, xf_2), (xn_3, xf_3)$ and (xn_4, xf_4) in this position. Then it can be linearly shifted by one or two pixels as unit step in the projected pixel plane. The methods of dividing area for each laser stripe of left and right images are the same. After dividing all the laser stripes area of left and right images, stereo matching and reconstruction work can be performed within the fixed areas.

3. Experiment and Analysis

3.1. Robust Test

In order to verify the robustness of the laser stripe projected by MEMS scanning mirror, the laser stripe images captured by cameras are investigated. Firstly, two laser stripes, one is projected by traditional linear laser device and another by a MEMS mirror, are simultaneously projected onto a flat surface. The captured image is shown in Figure 7a, wherein the left side is a traditional laser stripe and the right side is a MEMS projection laser stripe. The pixels in the up, middle and bottom cross-sections of the captured image are drawn in Figure 7b. It can be seen that the traditional laser stripe image has more speckle noise and the width and intensity distribution of laser stripe are not even, while the laser stripe projected by MEMS mirror is more even and finer, which is more helpful to improve the measure accuracy.

In addition, in order to test the depth of field of the MEMS mirror, the MEMS and DLP project the same size spot. The camera is aligned with the light exit of the MEMS scanning mirror and the DLP. By changing the distance between the camera and the light exit, the relationship between the spot size and the projection distance is tested. To prevent interference from other ambient light, the experiment is conducted in the darkroom. The size of the MEMS mirror spot and the size of the DLP spot at different projection distances are shown in Figure 8. The solid dot is the measurement value of the spot diameter and the curve is the relationship between the fitted spot size and the projection distance.

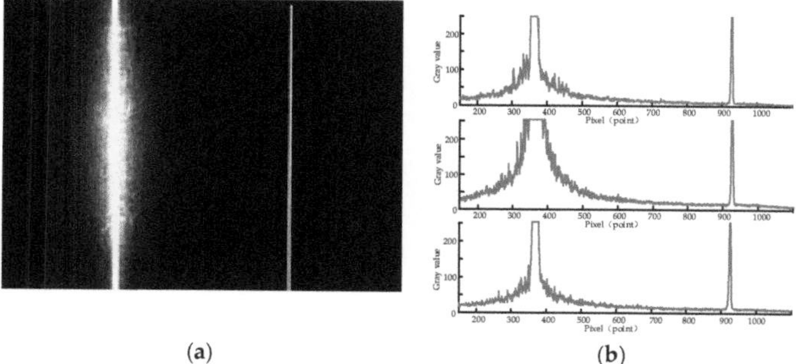

(a) (b)

Figure 7. Laser stripe contrast: (**a**) Two laser stripes captured in the same field of view; and (**b**) Gray distribution of different rows of pixels.

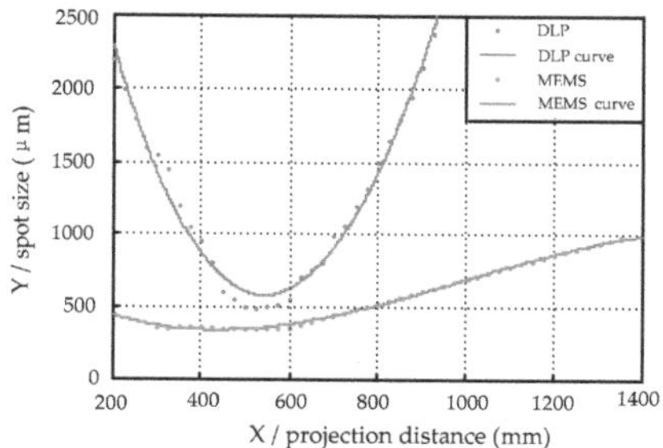

Figure 8. Depth of field comparison between Digital Light Processing (DLP) and MEMS scanning mirror.

It can be seen from the experiment that the spot projected by MEMS mirror is focused at 500 mm after passing through the lens, the spot size is about 300 μm and the spot diameter is kept within 0.5 mm within the 300 mm from the focus point. The red curve is the depth of field simulation of MEMS scanning mirror. The spot projected by DLP is also focused at 500 mm and the spot quickly blurs after deviating from the focus position. The blue curve is the depth of field simulation of DLP technology. The experiment results show that the spot size of the MEMS scanning mirror is smaller than DLP at the focus position and the spot diameter changes little within the measurement area. This experiment verifies the advantages of the MEMS scanning mirror with a large depth of field. In contrast, the DLP projection has a smaller depth of field and is sensitive to different reflectivity of object surface and to ambient light, which greatly reduce the accuracy of results. Thus DLP could not replace laser when using line structured light method in a complex environment. Therefore, DLP is generally used only as a fringe projection for full-field scanning measurement and the scanning time is a few seconds. The above experiments proved that the laser stripes projected by MEMS scanning mirror are robust and more suitable for line structured light measurement.

3.2. Reconstruction of A Face Plaster Model

A human-face plaster model was reconstructed by the proposed method and traditional laser-based 3D reconstruction method. The resolution of the MEMS scanning mirror is 1280 × 720 pixels. If the object is scanned pixel by pixel in the horizontal direction with a single laser stripe, the number of the laser stripe images needed to be captured is 1280. According to Equation (3), the most minimum pixel pitch for MEMS scanning mirror can be calculated as 285 pixels. Since the number of projected laser stripes must be an integer, the number of laser stripes is selected to be 4 for scanning and reconstruction. In this situation, only 320 images can implement the measurement of the face plaster model. The frame rate of the camera is 30 frames/s and the frame rate of the MEMS scanning mirror is 50 frames/s. The maximum frame rate of the system 30 frames/s is selected and the time taken for the field measurement is only 10.6 s. Then the laser stripe centerline is extracted by the gravity center extraction method [35] and the three-dimensional reconstruction of the model is implemented by binocular stereo matching method based on a fixed area.

Figure 9 shows the results of reconstruction by traditional laser scanning method and the MLSSP. Due to the large amount of speckle noise and the unevenness of the brightness of the conventional laser stripe, the images captured by cameras need filtering and removing noises before 3D reconstruction. Otherwise, a large number of mismatched points will appear in the stereo matching, which can reduce the accuracy of 3D reconstruction. Figure 9a shows the result of three-dimensional reconstruction by the conventional laser stipe scanning method after noise reduction. The texture features of the mouth and eyes are smoothed and errors may occur on the left face. For the MEMS scanning mirror, no noise reduction is required before 3D reconstruction. More importantly, no mechanical motion device is needed and it only takes 15 s to implement the scanning and 3D reconstruction. Compared with traditional methods, the measurement time by MLSSP is greatly reduced. Figure 9b is the 3D reconstruction model by MLSSP. The facial model is reconstructed very clearly and the surface curvature changes such as the nose and eyes are also clearly reconstructed. It can be seen that the 3D reconstruction model has higher accuracy by MLSSP contrasting to traditional laser-based 3D reconstruction method.

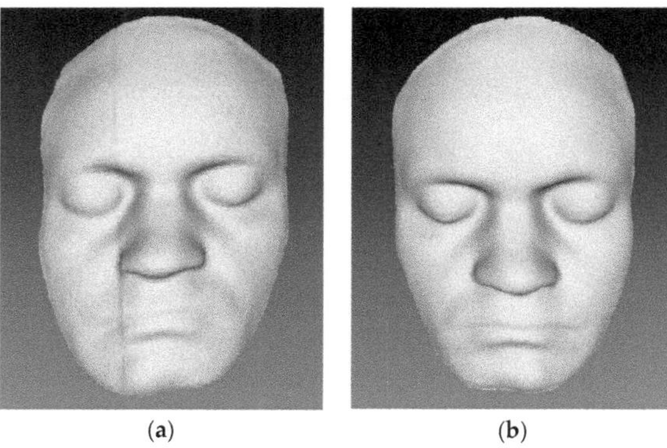

Figure 9. Three Dimensional reconstruction results contrast: (**a**) Traditional laser scanning method; and (**b**) MLSSP.

3.3. Accuracy Test

The precision of the system should be judged by measuring the standard objects with known size. As shown in Figure 10a two gauge blocks are stacked to form a standard step. The upper ceramic gauge block is 5 mm thick with tolerance of ± 6 μm. The three-dimensional model of gauge block above is reconstructed by the MLSSP as shown in Figure 10b. The red plane is the result of the reconstruction of the ceramic gauge above and the blue is the result of the reconstruction of the ceramic gauge below. The reconstructed profiles appear to be correct. To further check the measurement accuracy, the cross section of the recovered profiles with $x = 30$mm is plotted in Figure 10c. It is detected that the step height is around 5 mm. The measurement size of the gauge block above is subtracted by the true size of 5 mm to obtain an error distribution of all points as shown in Figure 11. The different colors represent the different errors of the upper surface of the standard gauge block. The maximum residual of the step size can be obtained as (/mm)

$$e_{\max} = \max(h_i - h_{real}) \tag{4}$$

where h_i is the measurement size of the ith pixel and h_{real} is the true size of 5 mm. If e_i is the measurement deviation of the ith pixel and $\bar{e}$ is the value of average deviation, the standard deviation is calculated by the following equation (/mm)

$$\sigma = \sqrt{\frac{1}{N}\sum_{i=1}^{N}(e_i - \bar{e})^2} \tag{5}$$

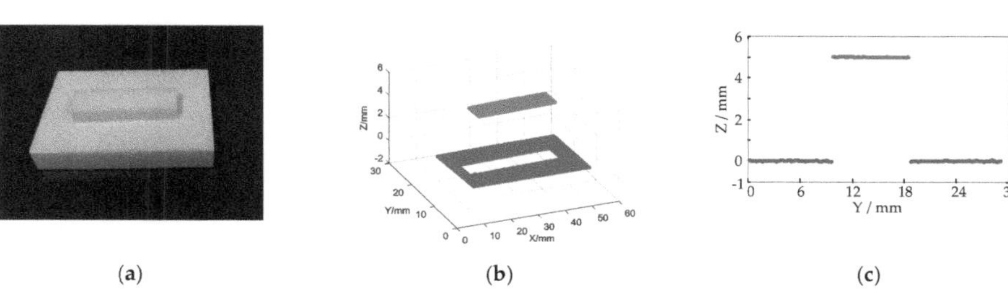

(a) (b) (c)

Figure 10. (**a**) Standard gauge block placed on a standard plane; (**b**) Three-dimensional model; (**c**) Cross section of the reconstructed 3D profile with $x = 30$ mm.

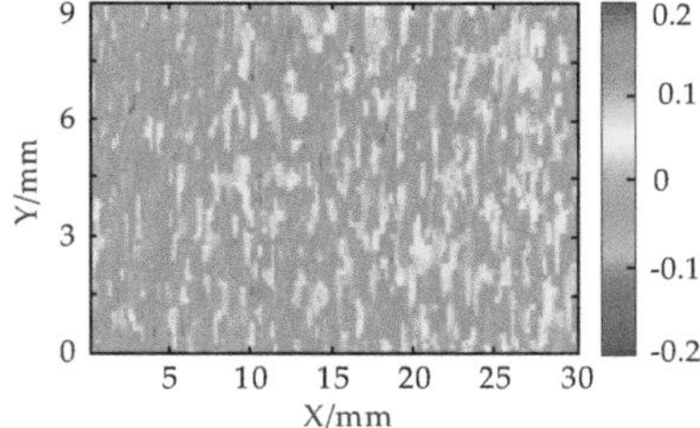

Figure 11. Measurement error distribution of standard gauge block upper surface.

According to measurement error distribution of standard gauge block upper surface, the result of $e_{\max}$ is 0.1428 mm and the standard deviation of σ is 0.0535 mm.

The image processing methods in the experiment use the gravity center extraction method [35] and linear interpolation stereo matching. However, the methods are not focal of my research, which can be replaced by other more accurate method to improve the accuracy of 3D reconstruction. In addition, the accuracy of the method is also affected by the surface accuracy of the ceramic block itself, the systematic errors and the external disturbances, which can be further reduced during the experiment. It can be seen that the measurement system of the MLSSP has excellent three-dimensional measurement performance and can provide powerful technological support for 3D inspection, reverse engineering and rapid manufacturing represented by 3D printing.

4. Conclusions

In summary, a robust multiple laser stripe scanning method for three-dimensional measurement is proposed in this work. It is based on programmable projection and fast scanning of MEMS scanning mirror. The test results show that the laser stripes projected by MEMS scanning mirror are more even and finer than traditional laser stripe. MEMS scanning mirror is more suitable for the line-structured light measurement than DLP and traditional laser. In addition, the number, period and scanning direction of multiple laser stripes projected by MEMS scanning mirror can be adjusted by calculating and programming. The MLSSP can implement the automatic scanning without the need of mechanical motion devices, which eliminates vibration noise caused by mechanical motion. Moreover, the number of the pictures needed to be captured is reduced to $1/N$. It greatly improves the efficiency of online 3D measurement. Finally, the reliability tests are conducted, including reconstruction of a face plaster model and accuracy test. The human-face plaster model is reconstructed clearly including the surface curvature changes of mouth and eyes. Accuracy test shows that standard deviation is 0.0535 mm and the accuracy of 3D reconstruction can be further improved.

Author Contributions: Conceptualization, G.H. and X.Z.; Data curation, G.H. and G.W.; Methodology, G.H. and X.Z.; Software, G.H., G.Z. and D.L.; Validation, G.H. and X.Z.; Visualization, G.H. and G.Z.; Writing–original draft, G.H.; Writing–review & editing, X.Z. and C.Z.

Funding: This research was supported by the National Science and Technology Major project (Grant No. 2017ZX04002001) and the China Postdoctoral Science Foundation (Grant No. 2018M633439).

Conflicts of Interest: The authors declare no conflict of interest.

References

1. Gong, Z.; Sun, J.; Zhang, G. Dynamic structured-light measurement for wheel diameter based on the cycloid constraint. *Appl. Opt.* **2016**, *55*, 198–207. [CrossRef]
2. Usamentiaga, R.; Molleda, J.; Garcia, D.F. Structured-Light Sensor Using Two Laser Stripes for 3D Reconstruction without Vibrations. *Sensors* **2014**, *14*, 20041–20063. [CrossRef]
3. Nguyen, H.; Nguyen, D.; Wang, Z.; Kieu, H.; Le, M. Real-time, high-accuracy 3D imaging and shape measurement. *Appl. Opt.* **2015**, *54*, 9–17. [CrossRef]
4. Lochner, S.J.; Huissoon, J.P.; Bedi, S.S. Development of a patient-specific anatomical foot model from structured light scan data. *Comput. Methods Biomech. Biomed. Eng.* **2012**, *17*, 1198–1205. [CrossRef]
5. Feng, S.; Zhang, Y.; Chen, Q.; Zuo, C.; Li, R.; Shen, G. General solution for high dynamic range three-dimensional shape measurement using the fringe projection technique. *Opt. Lasers Eng.* **2014**, *59*, 56–71. [CrossRef]
6. Trinderup, C.H.; Kim, Y.H.B. Fresh meat color evaluation using a structured light imaging system. *Food Res. Int.* **2015**, *71*, 100–107. [CrossRef]
7. Verdú, S.; Ivorra, E.; Sánchez, A.J.; Barat, J.M.; Grau, R. Relationship between fermentation behavior, measured with a 3D vision structured light technique and the internal structure of bread. *Food Eng.* **2015**, *146*, 227–233. [CrossRef]
8. He, L.; Wu, S.; Wu, C. Robust laser stripe extraction for three-dimensional reconstruction based on a cross-structured light sensor. *Appl. Opt.* **2017**, *56*, 823–832. [CrossRef]
9. Li, B.; Gibson, J.; Middendorf, J.; Wang, Y.; Zhang, S. Comparison between LCOS projector and DLP projector in generating digital sinusoidal fringe patterns. In Proceedings of the 2013 Conference on Dimensional Optical Metrology and Inspection for Practical Applications, San Diego, CA, USA, 25–26 August 2013.
10. Kannegulla, A.; Shams, M.I.B.; Liu, L.; Cheng, L. Photo-induced spatial modulation of THz waves: Opportunities and limitations. *Opt. Express* **2015**, *23*, 32098–32112. [CrossRef] [PubMed]
11. Huang, Y.; Pan, J. High contrast ratio and compact-sized prism for DLP projection system. *Opt. Express* **2014**, *22*, 17016–17029. [CrossRef] [PubMed]
12. Zhang, S.; van der Weide, D.; Oliver, J. Superfast phase-shifting method for 3-D shape measurement. *Opt. Express* **2010**, *18*, 9684–9689. [CrossRef]
13. Zuo, C.; Chen, Q.; Gu, G.; Feng, S.; Feng, F. High-speed three-dimensional profilometry for multiple objects with complex shapes. *Opt. Express* **2012**, *20*, 19493–19510. [CrossRef]
14. Zhou, X.; Yang, T.; Zou, H.; Zhao, H. Multivariate empirical mode decomposition approach for adaptive denoising of fringe patterns. *Opt. Lett.* **2012**, *37*, 1904–1906. [CrossRef] [PubMed]
15. Wang, Y.; Zhang, J.; Luo, B. High dynamic range 3D measurement based on spectral modulation and hyperspectral imaging. *Opt. Express* **2018**, *26*, 34442–34450. [CrossRef]
16. Sun, Q.; Chen, J.; Li, C. A robust method to extract a laser stripe centre based on grey level moment. *Opt. Lasers Eng.* **2014**, *59*, 56–71. [CrossRef]
17. Yin, X.Q.; Tao, W.; Feng, Y.Y.; Gao, Q.; He, Q.Z.; Zhao, H. Laser stripe extraction method in industrial environments utilizing self-adaptive convolution technique. *Appl. Opt.* **2017**, *56*, 2653–2660. [CrossRef]

18. Usamentiaga, R.; Garcia, D.; Molleda, J.; Bulnes, F.; Bonet, G. Vibrations in steel strips: Effects on flatness measurement and filtering. *IEEE Trans. Ind. Appl.* **2014**, *50*, 3103–3112. [CrossRef]
19. Petersen, K.E. Silicon torsional scanning mirror. *IBM J. Res. Dev.* **1980**, *24*, 631–637. [CrossRef]
20. Takashima, Y.; Hellman, B.; Rodriguez, J.; Chen, G.; Smith, B.; Gin, A.; Espinoza, A.; Winkler, P.; Perl, C.; Luo, C. MEMS-based Imaging LIDAR. In Proceedings of the Light, Energy and the Environment Congress 2018 in Sentosa Island, Singapore, 5–8 November 2018.
21. Tan, Y.; Dong, R. Nonlinear Model Based Control of MEMS Micro-Mirror. In Proceedings of the 2018 14th IEEE/ASME International Conference on Mechatronic and Embedded Systems and Applications (MESA), Oulu, Finland, 2–4 July 2018.
22. Sun, H.B.; Kawata, S. Two-Photon Laser Precision Microfabrication and Its Applications to Micro–Nano Devices and Systems. *J. Lightw. Technol.* **2003**, *21*, 624–633.
23. Liu, J.; Wang, J.; Wang, Y.; Tian, D.; Zheng, Q.; Lin, X.; Wang, L.; Yang, Q. Research on the compensation of laser launch optics to improve the performance of the LGS spot. *Appl. Opt.* **2018**, *57*, 648–651. [CrossRef]
24. Qi, W.; Chen, Q.; Guo, H.; XieD, H.; Xi, L. Miniaturized Optical Resolution Photoacoustic Microscope Based on a Microelectromechanical Systems Scanning Mirror. *Micromachines* **2018**, *9*, 288. [CrossRef] [PubMed]
25. Li, F.; Zhou, P.; Wang, T.; He, J.; Yu, H.; Shen, W. A Large-Size MEMS Scanning Mirror for Speckle Reduction Application. *Micromachines* **2017**, *8*, 140. [CrossRef]
26. Isamoto, K.; Totsuka, K.; Suzuki, T.; Sakai, T.; Morosawa, A.; Chong, C.; Fujita, H.; Toshiyoshi, H. A high speed MEMS scanner for 140-kHz SS-OCT. In Proceedings of the 2018 International Conference on Optical MEMS and Nanophotonics(OMN), Lausanne, Switzerland, 29 July–2 August 2018.
27. Isamoto, K.; Totsuka, K.; Sakai, T.; Suzuki, T.; Morosawa, A.; Chong, C.; Fujita, H.; Toshiyoshi, H. High speed MEMS scanner based swept source laser for SS-OCT. *IEEE J. Trans. Sens. Micromach. MEMS Packag. Microfabr. Technol.* **2012**, *132*, 254–260. [CrossRef]
28. Niclass, C.; Ito, K.; Soga, M.; Matsubara, H.; Aoyagi, I.; Kato, S.; Kagami, M. Design and characterization of a 256 × 64-pixel single-photon imager in CMOS for a MEMS-based laser scanning time-of-flight sensor. *Opt. Express* **2012**, *20*, 11863–11881. [CrossRef]
29. Liu, L.; Xie, H. 3-D Confocal Laser Scanning Microscopy Based on a Full-MEMS Scanning System. *IEEE Photonics Technol. Lett.* **2013**, *25*, 1478–1480. [CrossRef]
30. Pan, Y.; Xie, H.; Fedder, G.K. Endoscopic optical coherence tomography based on a microelectromechanical mirror. *Opt. Lett.* **2001**, *26*, 1966–1968. [CrossRef]
31. Miyajima, H.; Asaoka, N.; Isokawa, T.; Ogata, M.; Aoki, Y.; Imai, M.; Fujimori, O.; Katashiro, M.; Matsumoto, K. A MEMS electromagnetic optical scanner for a commercial confocal laser scanning microscope. *J. Microelectromech.* **2003**, *12*, 243–251. [CrossRef]
32. Raboud, D.; Barras, T.; Conte, F.L.; Fabre, L.; Kilcher, L.; Kechana, F.; Abelé, N.; Kayal, M. MEMS based color-VGA micro-projector system. *Procedia Eng.* **2010**, *5*, 260–263. [CrossRef]
33. Khayatzadeh, R.; Civitci, F.; Ferhanoglu, O.; Urey, H. Scanning fiber microdisplay: design, implementation, and comparison to MEMS mirror-based scanning displays. *Opt. Express* **2018**, *26*, 5576–5590. [CrossRef]
34. Tanguy, Q.A.; Bargiel, S.; Xie, H.; Passilly, N.; Barthès, M.; Gaiffe, O.; Rutkowski, J.; Lutz, P.; Gorecki, C. Design and fabrication of a 2-Axis Electrothermal MEMS Micro-scanner for optical Coherence Tomography. *Micromachines* **2017**, *8*, 146. [CrossRef]
35. Zhao, B.H.; Wang, B.X.; Zhang, J.; Luo, X.Z. Extraction of laser stripe center on rough metal surface. *Opt. Precis. Eng.* **2011**, *19*, 2138–2145. [CrossRef]

Article

Hybrid 3D Shape Measurement Using the MEMS Scanning Micromirror

Tao Yang [1], Guanliang Zhang [1], Huanhuan Li [1] and Xiang Zhou [1,2,*]

[1] School of Mechanical Engineering, Xi'an Jiaotong University, Xi'an 710049, Shaanxi, China; xjtu.yangtao@gmail.com (T.Y.); gl-zhang@foxmail.com (G.Z.); xjtu.vivianli@gmail.com (H.L.)
[2] School of Food Equipment Engineering and Science, Xi'an Jiaotong University, Xi'an 710049, Shaanxi, China
* Correspondence: zhouxiang@mail.xjtu.edu.cn

Received: 10 December 2018; Accepted: 7 January 2019; Published: 11 January 2019

Abstract: A surface with large reflection variations represents one of the biggest challenges for optical 3D shape measurement. In this work, we propose an alternative hybrid 3D shape measurement approach, which combines the high accuracy of fringe projection profilometry (FPP) with the robustness of laser stripe scanning (LSS). To integrate these two technologies into one system, first, we developed a biaxial Microelectromechanical Systems (MEMS) scanning micromirror projection system. In this system, a shaped laser beam serves as a light source. The MEMS micromirror projects the laser beam onto the object surface. Different patterns are produced by controlling the laser source and micromirror jointly. Second, a quality wised algorithm is delivered to develop a hybrid measurement scheme. FPP is applied to obtain the main 3D information. Then, LSS helps to reconstruct the missing depth guided by the quality map. After this, the data fusion algorithm is used to merge and output complete measurement results. Finally, our experiments show significant improvement in the accuracy and robustness of measuring a surface with large reflection variations. In the experimental instance, the accuracy of the proposed method is improved by 0.0278 mm and the integrity is improved by 83.55%.

Keywords: MEMS scanning micromirror; fringe projection; laser stripe scanning; quality map; large reflection variations

---

1. Introduction

Three-dimensional (3D) shape information can be widely used in human–computer interaction [1,2], biometric identification [3,4], robot vision [5,6], virtual/augmented reality [7,8], industry [9] and other fields. As a result, 3D shape measurement attracts a lot of attention in the community of computer science and instrument science.

Fringe projection profilometry (FPP) is considered one of the most popular approaches because of the advantages of non-contact operation, high accuracy and full-field acquisition [10,11]. In FPP, sinusoidal fringes are projected onto a measuring surface by using a digital projector. Meanwhile, the observation pattern images are obtained from another angle using a camera. We can decode the height of the surface by analyzing the distortion of the observation fringe patterns [9,12,13]. However, FPP assumes that the measuring surface exhibits a diffuse reflection and usually considers low-reflective (dark) and highlighted (specular reflection) areas as outliers. These regions can completely block any fringe patterns, which results in the loss of depth information [14–16]. To address this problem, some solutions are presented. In [17], Salahieh et al. propose a multi-polarization fringe projection (MPFP) imaging technique that handles high dynamic range (HDR) objects by selecting the proper polarized channel measurements. A similar polarization solution is also adopted in [18]. On the other hand, Liu et al. demonstrate the use of a dual-camera FPP system, which can also be considered as two camera-projector monocular systems. By viewing from different angles, these highly specular and dark pixels, which are missing from binocular reconstruction, can be filled in [19]. In addition, Jiang et al. [20] present using 180-degree phase-shifted (or inverted) fringe patterns to reduce the measurement error for high-contrast surfaces reconstruction. Some other researchers have attempted to adjust the parameters of the camera and projector to handle the surface with large reflection variations. Lin et al. [21,22] suggest adjusting the maximum input gray level of projecting image globally, while Chen et al. [23,24] proposed adjusting projecting images pixel-by-pixel. In [16,25], the author proposes projecting a set of fringe images that are captured with different exposures. For [16,21–25], the reflection of the surface needs to be calibrated first. Then, by fusing these images captured with different parameters, a new fringe pattern with fewer saturated regions can be obtained. Although Jiang et al. [16,20–25] improve the performance without adding any extra equipment, they need to project or capture a lot of images when the measuring surface has very complex reflection variations. On the other hand, these approaches still use an FPP principle, which can be ineffective for extreme reflection areas.

Laser stripe scanning (LSS) [26,27] is another kind of structured light approach, which shares the same triangulation [28] measurement principle with FPP. The difference is that LSS applies a scanning stripe pattern instead of a fringe pattern. As LSS just needs to extract the stripe in the observation images, it results in very high robustness [29–31]. Therefore, LSS is widely used in the 3D shape measurement industry [32]. However, it is expensive to obtain a very high accuracy in LSS, which is mainly determined by the width of the stripe pattern and the resolution of the camera. It is reasonable to consider whether we could use the same hardware to set up an LSS and an FPP system to handle different surface reflection. Generally, because of the limited projection depth of the field, the answer is no. In FPP, Digital Light Processing (DLP) or a Liquid Crystal On Silicon (LCOS) projector is used to project fringe patterns on the measuring surface [33,34]. Those projection techniques can only produce sharp images in the focal plane. If a stripe is projected, it will be severely blurred on the defocused plane. This means that a commercial projector cannot work for LSS. In LSS, a laser stripe projector is adopted, and the object surface is scanned by moving the object or measurement system. A typical laser stripe projector employs a laser beam as the light source, a cylindrical lens is used to scatter the laser beam into a stripe. Therefore, it can't produce a fringe pattern generally. However, by using Microelectromechanical Systems (MEMS) projection technology, it is possible to generate a stripe pattern and fringe pattern with the same hardware. In MEMS projection [35], a biaxial (or single axial) MEMS micromirror [36–38] is applied to scan a laser beam point by point (row by row for single axial MEMS scanner with a cylindrical lens) to produce the projection pattern. In Ref. [39], the authors set up a compact 3D shape measurement system with a single axial MEMS micromirror. As only the FPP principle is used in this work, they still cannot measure the surface with large reflection variations.

In this paper, we propose a hybrid 3D shape measurement approach, which employs FPP and LSS in the same system by applying a biaxial MEMS scanning micromirror to generate the fringe pattern and scanning laser stripe with the same setup. By doing so, the proposed method can handle the surface that has large reflection variations with high accuracy and high robustness.

2. Principles

2.1. Principle of 3D Shape Measurement with FPP and LSS

2.1.1. Principle of Fringe Projection Profilometry

As shown in Figure 1a, a typical FPP measurement system consists of a digital projector and a digital camera [10]. The light axis (EpO) of the projector intersects the light axis (EcO) of the Charge-coupled Device (CCD) camera at Point O in the reference plane (along the x-axis). The distance between the two optical centers is d, and the distance between the camera and the reference plane is l. Point D is an arbitrary point on the object's surface with a height of h. Points A and C are the intersections of the light paths of the projector and the camera, respectively, with the reference axis. Compared with projecting a sinusoidal fringe pattern onto the reference plane, when an object is placed on the reference plane, the fringe pattern captured by the CCD will be distorted by the object's height. The modulated phase difference will have a relationship with the true height h, as given by Equation (1), where the inference process can be found in [12]:

$$h_{FPP}(x,y) = \frac{l\Delta\phi(x,y)}{2\pi f_0 d}, \tag{1}$$

where f_0 is the frequency of the fringe pattern projection. $\Delta\phi$ represents the phase difference between Point D and Point A, which is equal to the phase difference between Point A and Point C in the reference plane.

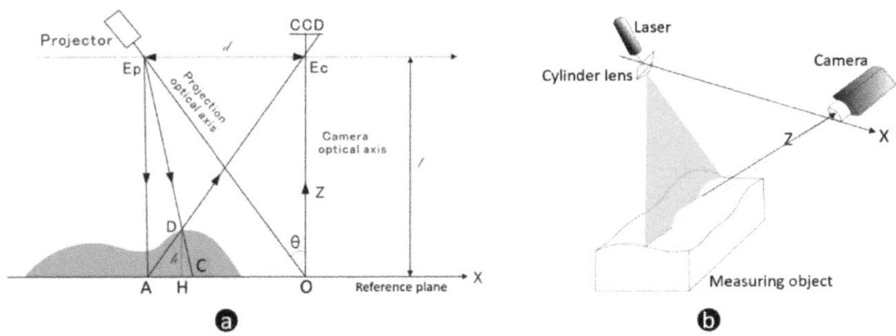

Figure 1. Schematic drawings of fringe projection and laser stripe scanning. (**a**) basic light path of Fringe projection profilometry system; (**b**) schematic diagram of laser stripe scanning system.

2.1.2. Principle of Laser Stripe Scanning

Laser stripe scanning is based on active laser-triangulation (Figure 1b). In LSS, a laser stripe, created by a dot laser and then scattered by a cylinder lens, is projected onto the measuring object surface and the reflection light is observed under the triangulation angle with a digital camera [26]. Changing the distance between the object and measurement system results in a movement of laser stripe's position in the x-direction observed with the z-direction. This position is calculated by extracting the laser stripe center. LSS thus delivers a height distribution of the object. In most cases, industry applications need to make full-field measurements where a highly accurate moving part is introduced. The moving part changes the position between the system and the object, so that the laser stripe sweeps across the surface of the object to obtain full field height distribution. Similar to FPP, when an object is placed on the reference plane, the laser stripe delivers Δx movement, and Equation (2) shows the relationship between Δx and height $h_{LSS}\,(x,y)$:

$$h_{LSS}\,(x,y) = \frac{d\Delta x}{l}, \tag{2}$$

where d is the distance between the laser and the camera. l is measurement distance.

2.2. Hybrid 3D Shape Measurement System

2.2.1. Biaxial MEMS Micromirror-Based Pattern Projection

The biaxial MEMS micromirror-based pattern projection system is the foundation of our pipeline, which can produce both a stripe pattern and fringe pattern. Figure 2 shows the basic layout of our MEMS pattern projection system. There is a single model laser diode (LD) which served as the light source. Meanwhile, near the LD, an aspheric lens is placed to adjust the focus of the laser beam and shape the beam. Before the laser beam is relayed onto the biaxial MEMS micromirror, there is an aperture to remove the stray light. A biaxial electromagnetic actuation MEMS micromirror working on raster scan mode reflects the light source to the object surface to produce a different 2D pattern. Both of the fast and slow axes rotate reciprocally driven by the current signal, which makes the micromirror scan the laser beam row by row. At the same time, the intensity of the laser is modulated under the synchronization of the sync signal. Figure 3a,b illustrate the controlling signals for fringe pattern and stripe pattern projection. H and V are the horizontal and vertical driven signals of the MEMS micromirror, respectively. *Sync* is the row sync signal. Additionally, *Laser* modulates the LD to produce different intensity. It should be noted that both the horizontal and vertical axes just operate in resonant vibration mode, no matter whether the fringe pattern or stripe pattern is projected. As shown in Figure 3, H and V are always sinusoidal waveforms. This kind of variable-speed scanning introduces distortion in pattern projection, which can be considered as a phase error resulting in a systematic error. To remove the distortion, pre-correction is performed on *Laser* generally, where more detailed information can be found in [35].

Different from the pixel array based projection technique, a MEMS micromirror based pattern projection system produces different patterns by scanning the laser beam two-dimensionally. This makes it possible to project fringe pattern and stripe pattern with the same hardware. Meanwhile, due to a laser source having better linearity than the light-emitting diode (LED) source, no gamma correction [40] is required in FPP with proposed pipeline, which brings additional benefits.

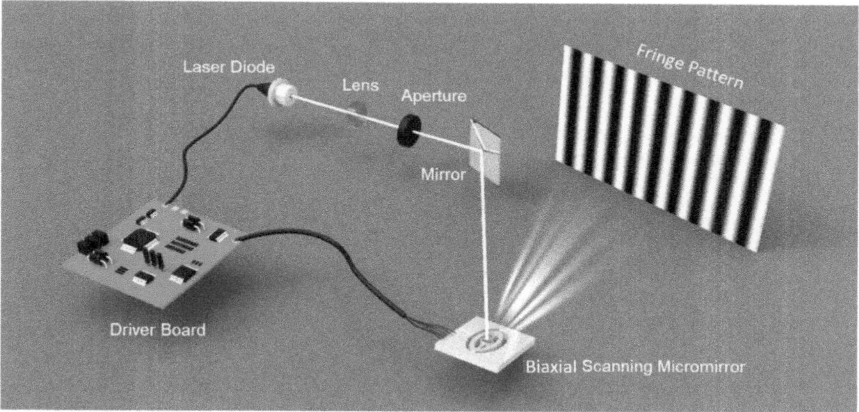

Figure 2. Schematic diagram of the biaxial Microelectromechanical Systems (MEMS)-based fringe pattern projection system.

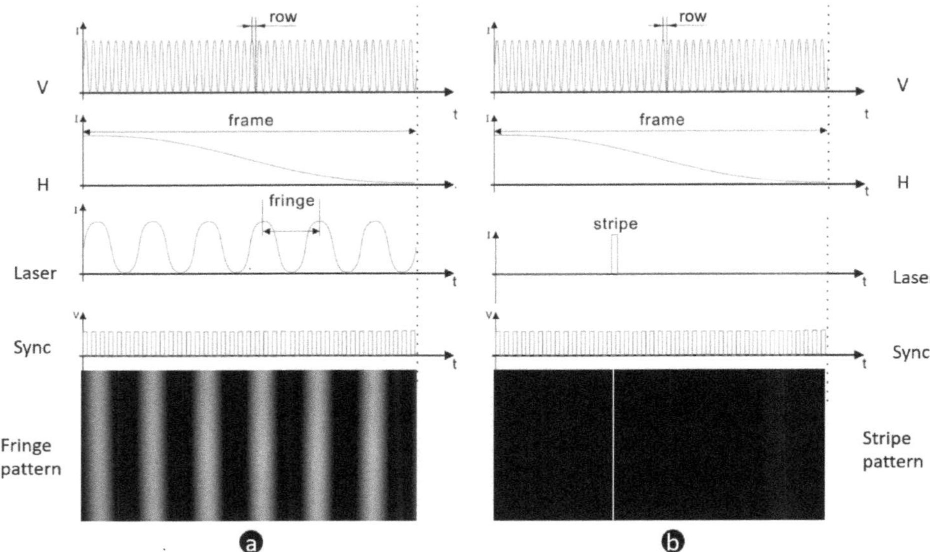

Figure 3. Controlling signals of pattern projection. (**a**) controlling signal of fringe projection; (**b**) controlling signal of stripe projection. *H* and *V* are the horizontal and vertical driven signals of the MEMS micromirror. *Sync* is the row sync signal. *Laser* is the modulation signal of the laser diode (LD).

2.2.2. Quality Index in the Proposed Approach

Our system is significantly simplified by implementing FPP and LSS in the same system. Another benefit is that the data from FPP and LSS are based on the same coordinate system, which makes it much easier to align these two parts of the data. Before data fusion, we need to build an error model to evaluate the quality of measuring data, so that we can guide the process of data fusion. In this section, we define a quality index for our hybrid measurement approach.

In FPP, the phase of each point is calculated by the phase shift method. The fringe image obtained by the CCD can be described by

$$I_{ci}(x,y) = a_0(x,y) + b_{mod}(x,y)\cos(\Phi + 2\pi n_i/N_s) + n(x,y). \quad (3)$$

In Equation (3), $a_0(x,y)$ is the respective backgrounds, while $b_{mod}(x,y)$ is the respective modulation functions, also called the contrast. In addition, N_s is the number of steps of phase shifting, n_i is an integer, and $n(x,y)$ is the random noise.

In fact, we modulate the phase by changing the grayscale of the projection images. Here, we would like to discuss how the grayscale affects measurement accuracy. In the N-steps phase shifting method, the phase shifting noise caused by random noise is determined by N_s and the distribution of $n(x,y)$. We assume that this part noise $n_s \in [-N, N]$, as shown in Figure 4, where the complex plane represents the phase calculated by an imaginary part and a real part. If point P is a measurement point, the coordinates of Point P are $P(b_{mod}\cos\Phi, b_{mod}\sin\Phi)$. Due to the noise n_s, P will and change within the blue square (Figure 4), which has a side length of $2N$. If $n = (-N, N)$, then P will shift to Q and P will have a maximum phase error of $\varphi_{\max}$. If we assume that the phase of P is Φ, then the phase of Q can be given by

$$\Phi + \varphi = \arctan(\frac{b_{mod}\sin(\Phi) + N}{b_{mod}\cos(\Phi) - N}). \quad (4)$$

Thus, the phase difference is φ. Supposing that $k = \frac{N}{b_{mod}} \in [0, 1)$, then Equation (4) can be simplified as

$$\varphi = \arctan\left(\frac{\sin(\Phi) + k}{\cos(\Phi) - k}\right) - \Phi. \tag{5}$$

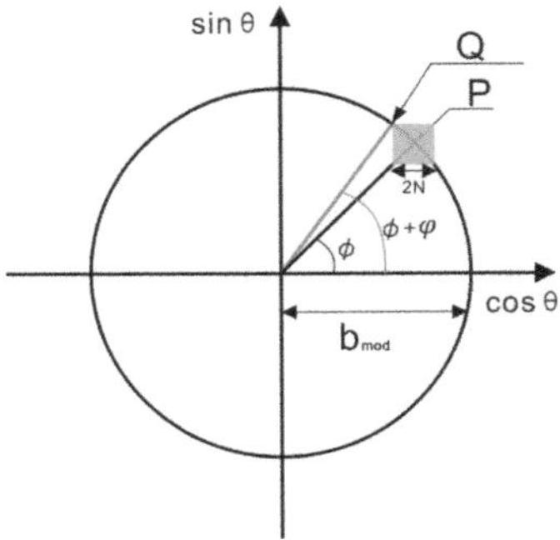

Figure 4. Error analysis in fringe projection profilometry (FPP).

From Equation (5), we know that the error of the FPP system is determined by k. This means that we can evaluate FPP depth data by b_{mod}.

In a fixed experiments setup, the noise in Equation (3) considered as a constant variable generally. Therefore, it can be simplified as Equation (6):

$$\begin{aligned} I_{ci}(x,y) &= A + b_{mod}\cos(\Phi + 2\pi n_i / N_s) \\ &= A + b_{mod}\cos(\Phi + \delta_i) \\ &= A + b_{mod}(\cos\Phi\cos\delta_i - \sin\phi\sin\delta_i) \\ &= A + B_c\cos\delta_i + B_s\sin\delta_i, \end{aligned} \tag{6}$$

where A is the combine of background $a_0(x,y)$, noise $n(x,y)$, and b_{mod} is the modulation in FPP:

$$\begin{cases} A = a_0(x,y) + n(x,y), \\ B_c = b_{mod}\cos\phi, \\ B_s = -b_{mod}\sin\phi, \\ \delta_i = i\frac{2\pi}{N_s}; i = 0, 1\cdots N_s - 1. \end{cases} \tag{7}$$

Based on the principle of trigonometric function orthogonality,

$$\begin{cases} B_c = \frac{2}{N_s}\sum I_i\cos\delta_i, \\ B_s = \frac{2}{N_s}\sum I_i\sin\delta_i. \end{cases} \tag{8}$$

Thus, we have

$$b_{mod} = \sqrt{{B_c}^2 + {B_s}^2}. \tag{9}$$

It can be known from Equation (5) that the quality of depth data is positively correlated with the modulation of fringe images. Therefore, we choose to conclude that modulation as the guiding quality index for data fusion in the proposed pipeline, where the quality index Q can be defined in Equation (10):

$$Q = \sqrt{\left(\frac{2}{N_s}\sum I_{ci}\cos\delta_i\right)^2 + \left(\frac{2}{N_s}\sum I_{ci}\sin\delta_i\right)^2}. \tag{10}$$

The region with high reliability, where the Q is above a threshold, will use the depth data from FPP. As a supplement, depth information that comes from LSS is used to fill in the other part, where the Q is not good enough. This allows us to keep the integrity with the optimal accuracy.

2.2.3. Hybrid 3D Shape Measurement Pipeline

Our Hybrid 3D shape measurement pipeline is shown in Figure 5. Both FPP and LSS are implemented with one system. Before measuring, we should first calibrate the camera and the projector. The calibration process provides the relationship between the height and distortion of structured Pattern [41]. FPP is adopted to obtain the quality map and depth map 1, while LSS is employed to obtain depth map 2. Generally, depth 1 has high accuracy but loses some depth information because of the extreme reflection. On the other hand, map 2 has lower accuracy but loses little depth information due to excellent robustness. Because map 1 and map 2 are naturally aligned, we can fill the final depth map by selecting a better part from map 1 and map 2 without any registration. The quality map shows where FPP works well; therefore, it is employed to guide the data fusion. When the quality index in quality map is above a threshold, the depth coming from FPP is considered high quality and will be used for the final depth map. Otherwise, LSS data will be adopted, and even both FPP and LSS data are available. Finally, the 3D surface is reconstructed from the refined depth map and calibration parameters. A carefully selected threshold is very important for proposed pipeline, and it will be presented in the next section.

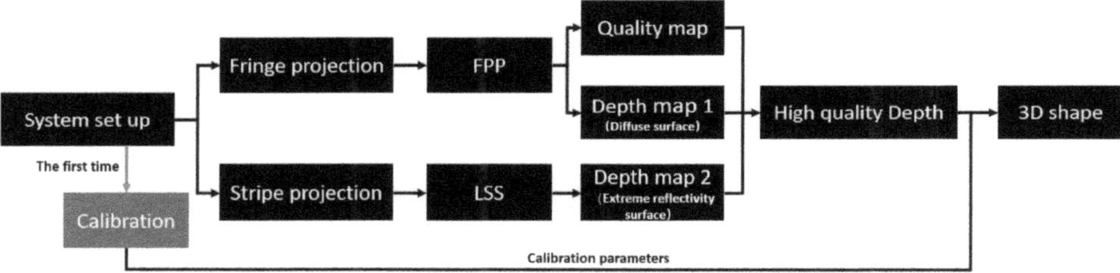

Figure 5. Flow chart of the proposed hybrid measurement approach.

3. Experiments

To implement the proposed hybrid measurement approach, we build our system as in Figure 2. Figure 6 illustrates our experimental setup. A MEMS pattern projector, driven by the driver board, produces both the fringe pattern and stripe pattern. In addition, a USB hub is adopted for communication and collection of the image data. A CCD camera with a 12 mm lens is used to capture the pattern images. The angle between the MEMS projector and the camera is set to 15 degrees, which can balance measuring resolution and coincidence of field of view. While the measuring distance is set to 500 mm, where the projector and CCD have the largest coincidence field of view and the best projection pattern quality. All these modules are mounted on an aluminum alloy casing with the overall size of 187 mm × 90 mm × 45 mm. More detailed parameters and description for these components are in Table 1. Both camera, USB hub and lenses are commercial products. The optical and mechanical part are designed by our own team. The driver board and MEMS scanner are designed and manufactured by the cooperation team. With all these setups, the hybrid 3D shape measurement system achieves a 0.07 mm measuring accuracy with a 286 mm × 176 mm field of view at an optimum working distance of 500 mm. Additionally, laser beam scanning has a larger depth of field than pixel array based projection. Hence, the depth of view of proposed system is mainly determined by the limitation of camera. It is about 100 mm in this experiment.

Table 1. Parameters and description for experimental components.

Items	Parameters and Description
Camera	Model: Charge-coupled Device (CCD), mono, global shutter. Resolution: 1920 × 1200. Max frames per second: 163
Lens for camera	Focal length: 12 mm
MEMS scanner	Model: 2D electromagnetic actuation. Fast axis: 18 kHz, ±16°. Slow axis: 0.5 kHz, ±10°
Laser diode	Wavelength: 650 nm. Power consumption: 320 mW
Lens for LD	Model: aspheric lens. Focal length: 4.51 mm. Aspheric coefficient: −0.925. Distance from LD: 4.55 mm. Distance from MEMS scanner: 33 mm
USB hub	USB3.0 × 5

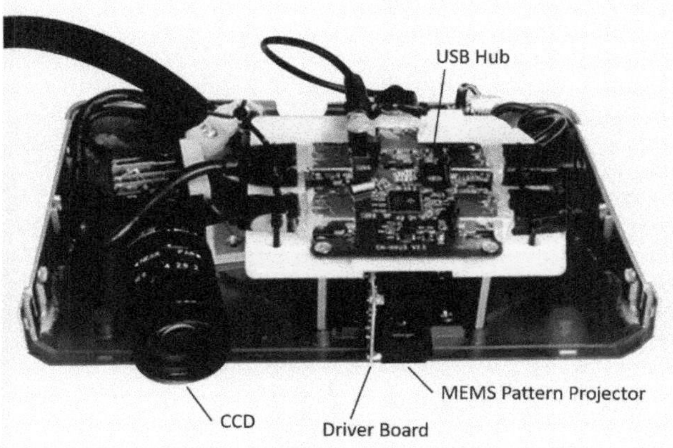

Figure 6. Hybrid 3D shape measurement setup.

3.1. Linearity Test of Proposed Pattern Projection System

In FPP, the nolinearity, which makes ideal sinusoidal waveforms nonsinusoidal, can significantly influence the performance [40]. Due to the use of a laser source, the proposed system has better linearity than a conventional LED based pattern projector. To verify the linearity of these two kinds of projectors, a commercial projector (coolux S3) is chosen as a comparison, which is an LED based DLP projector. Both projectors project pure white images with different gray levels. Then, an illuminometer is used to measure the illuminance with a fixed distance. Figure 7a,b illustrate the relationship between the projection gray value and illuminance of the proposed system and a commercial LED projector. As is shown in Figure 7, the proposed system has better linearity than a commercial LED projector. In fact, these two curves give luminescence characteristics of laser source and LED source. Therefore, gamma calibration, which is usually applied to eliminate the influence of nonlinearity in FPP [40], is no longer needed.

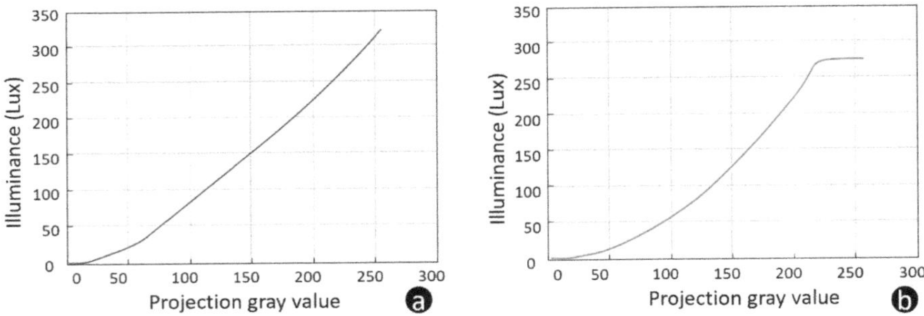

Figure 7. Linearity test. (**a**) proposed system; (**b**) commercial LED projector, item model: coolux S3.

3.2. Experiments on the Quality Index

In the previous section, we demonstrated the use of the quality index to evaluate depth information, and give the theoretical basis. In this section, we will verify the given theoretical basis and find the best threshold for data fusion by experiments. Here, we use a standard gauge block (30 mm × 50 mm × 8 mm) as the measuring object. The flatness of the block is less than 5 μm. The distribution of the depth value reflects the quality of depth data. In this work, we take the uncertainty of depth data as the evaluation of the measurement accuracy. Additionally, it is computed with Equation (11),

$$RMS_{error} = \sqrt{\frac{1}{N}\sum_{i=1}^{N}(e_i - \bar{e})^2}, \tag{11}$$

where e_i is the height of real measuring points and $\bar{e}$ is the height of the same position in the fitted plane by Equation (12) with the least squares method. N is the total number of real measuring points, and i stands for the index of a point:

$$A_{fit}x + B_{fit}y + C_{fit}z + D_{fit} = 0. \tag{12}$$

To find the relationship between fringe modulation and measurement accuracy, we set the modulation of the projected fringe pattern as different values to obtain the related uncertainty. Figure 8a is the standard gauge block. Figure 8b is the standard gauge block with the fringe pattern. Figure 8c illustrates the relationship between the modulation of the captured fringe pattern and measurement error. Obviously, lower modulation results in higher measurement error.

In Equation (2), the uncertainty of $h_{LSS}(x,y)$ is determined by the resolution of Δx due to the Δx being observed with a CCD camera. Thus, the resolution of Δx is a spatial distance, which is one pixel in the observation frame. It can be calculated as

$$\Delta x_{\min} = \frac{l\delta}{f}, \tag{13}$$

where l is the measurement distance, δ and f are the size of pixel and the focal length of the lens of observation camera. In these experimental settings, $l = 500$ mm, $\delta = 0.0064$ mm and $f = 12$ mm. Jointly with Equation (2), we have $\Delta h_{LSS}(x,y) = 0.0715$ mm. From Figure 8, when the modulation is above 32, FPP presents better accuracy than 0.0715 mm, which is the resolution of LSS. Therefore, the threshold for data fusion is chosen to be 32.

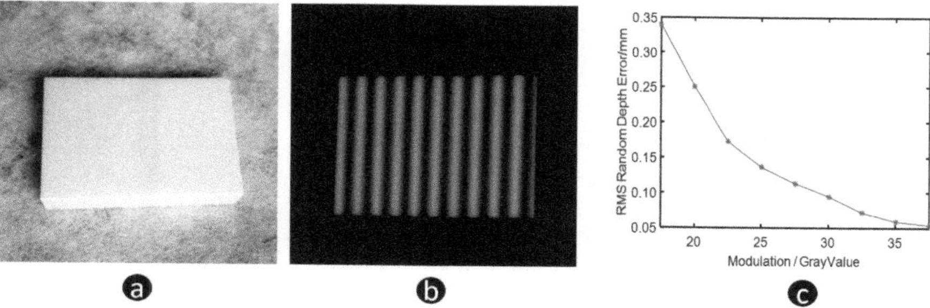

Figure 8. Depth reliability experiments. (**a**) standard gauge block; (**b**) standard gauge block with fringe pattern; and (**c**) the relationship between modulation and measurement error.

3.3. Hybrid 3D Shape Measurement

In this section, a black porcelain doll is chosen as our measuring object—see Figure 9a. For a black porcelain surface, the normal direction has a strong influence on the reflection. Weak reflection can be observed in most regions. When the normal direction is towards the camera, extremely strong reflection will be captured. This makes it difficult for FPP to work well. As a comparison, we paint the body of the doll with developer, which makes the body an ideal diffuse surface. Figure 9b is the quality map, where the threshold is 25. Figure 9c illustrates the fusion mask. Guided by the mask, we fuse the depth of FPP Figure 10a with the depth of LSS Figure 10b, and then we obtain the optimal 3D shape information Figure 10c. From Figure 10, we can find that FPP gives better accuracy (body of doll), but it cannot handle large reflectance variations (head of doll). LSS shows high robustness with different reflectance (head of doll) but lower accuracy (body of doll). The fusion data combines the advantages of FPP and LSS. To evaluate the measurement data quantitatively, a 3D measurement instrument (see Figure 11a) with a resolution of ± 5 μm, is adopted to offer the ground truth data (see Figure 11b). Because of the limitation of measurement range and efficiency, only the head of the doll is scanned. Then, the FPP reconstruction result and hybrid measurement result are registered with ground truth data to compute the 3D geometric error, respectively. As illustrated in Figure 12a, for FPP data, the root mean square (RMS) error is 0.0991 mm, and 0.0713 mm for hybrid measuring results in Figure 12b. In this case, integrity is computed to evaluate the robustness. For FPP, it is 54.48% taking hybrid results as the reference (100%). The experiments show that the proposed hybrid approach gets an improvement of 0.0278 mm in accuracy and 83.55% in integrity under the conditions described in this paper.

To verify the performance of the proposed approach, we chose several objects with large reflection variations that FPP cannot handle integrally. Figure 13a shows a plastic car model with shiny, dark and specular reflection regions. Figure 13b shows a metal surface. When the normal direction changes dramatically, it becomes hard to scan a metal surface. Figure 13c,d also show two difficult cases: stone material with very low reflection and a plastic surface with multiple reflectance and a large variation of normal directions. These results show the excellent performance of our approach.

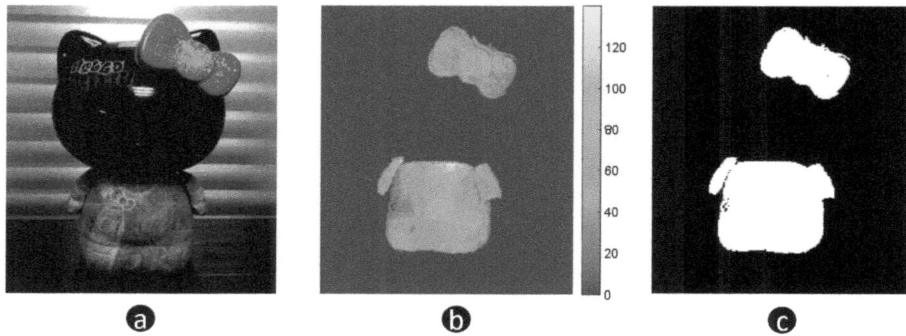

Figure 9. Measuring object and intermediate data. (**a**) measuring object; (**b**) quality map from FPP; (**c**) binary mask for data fusion.

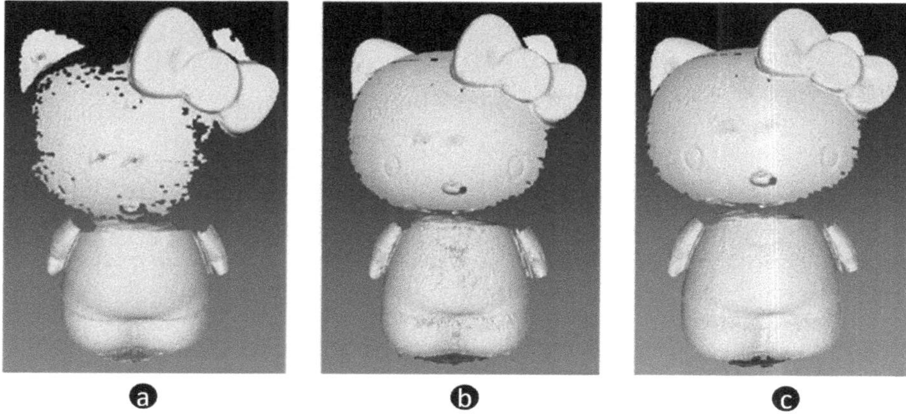

Figure 10. 3D reconstruction results. (**a**) reconstructing from FPP; (**b**) reconstructing from laser stripe scanning (LSS); (**c**) fusion results.

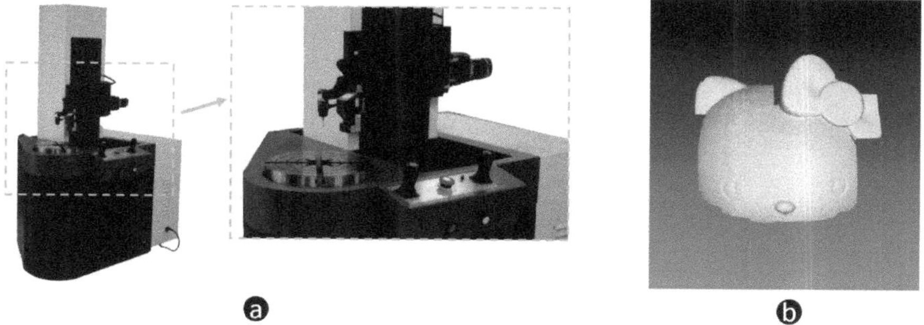

Figure 11. Ground truth data. (**a**) high accuracy 3D measurement equipment, resolution = ±5 μm; (**b**) high accuracy ground truth measurement result.

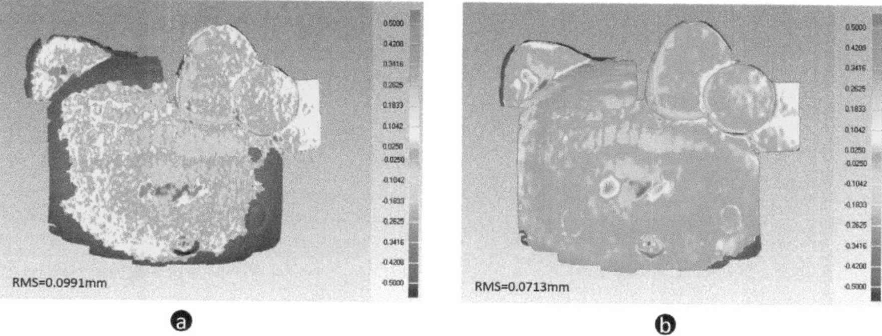

Figure 12. Accuracy evaluation. (**a**) comparison between FPP data and ground truth; (**b**) comparison between the data measured with proposed method and ground truth.

Figure 13. Experimental results of the proposed hybrid measurement approach. (**a**) plastic surface with shiny, dark and specular reflection regions; (**b**) metal surface, which results in shiny and dark regions from different perspectives of observation; (**c**) stone material with very low reflection; (**d**) plastic surface with multiple reflectance and large variation of normal directions.

4. Conclusions

In this paper, we have addressed the 3D shape measurement for large reflection variations with a hybrid approach. We proposed using a biaxial MEMS scanning micromirror and laser source to produce the fringe pattern and stripe pattern with the same hardware. Both FPP, which has the advantages of high accuracy and high efficiency, and LSS, which is one of the most robust methods, are employed to achieve a hybrid 3D shape measurement approach. Real experiments of different objects with large reflectance variations were carried out to verify the proposed method. The metal, plastic and stone materials with large reflection variations and large normal direction variations were reconstructed successfully, which shows the excellent performance of our method.

Author Contributions: Conceptualization, T.Y.; Data Curation, T.Y. and G.Z.; Formal Analysis, G.Z. and H.L.; Funding Acquisition, X.Z.; Methodology, T.Y. and G.Z.; Project Administration, T.Y.; Resources, H.L. and X.Z.; Software, G.Z.; Supervision, X.Z.; Visualization, T.Y.; Writing—Original Draft, T.Y., G.Z. and H.L.

Funding: This work was supported by the National Science and Technology Major project (No. 2015ZX04001003).

Conflicts of Interest: The authors declare no conflict of interest.

References

1. Rautaray, S.S.; Agrawal, A. Vision based hand gesture recognition for human computer interaction: A survey. *Artif. Intell. Rev.* **2015**, *43*, 1–54. [CrossRef]
2. Raheja, J.L.; Chandra, M.; Chaudhary, A. 3D gesture based real-time object selection and recognition. *Pattern Recognit. Lett.* **2018**, *115*, 14–19. [CrossRef]
3. Soltanpour, S.; Boufama, B.; Wu, Q.J. A survey of local feature methods for 3D face recognition. *Pattern Recognit.* **2017**, *72*, 391–406. [CrossRef]
4. Zulqarnain Gilani, S.; Mian, A. Learning From Millions of 3D Scans for Large-Scale 3D Face Recognition. In Proceedings of the IEEE Conference on Computer Vision and Pattern Recognition, Salt Lake City, UT, USA, 18–22 June 2018; pp. 1896–1905.
5. Faessler, M.; Fontana, F.; Forster, C.; Mueggler, E.; Pizzoli, M.; Scaramuzza, D. Autonomous, vision-based flight and live dense 3d mapping with a quadrotor micro aerial vehicle. *J. Field Robot.* **2016**, *33*, 431–450. [CrossRef]
6. Kim, H.; Leutenegger, S.; Davison, A.J. Real-time 3D reconstruction and 6-DoF tracking with an event camera. In Proceedings of the European Conference on Computer Vision, Amsterdam, The Netherlands, 8–16 October 2016; pp. 349–364.
7. Sra, M.; Garrido-Jurado, S.; Schmandt, C.; Maes, P. Procedurally generated virtual reality from 3D reconstructed physical space. In Proceedings of the 22nd ACM Conference on Virtual Reality Software and Technology, Munich, Germany, 2–4 November 2016; pp. 191–200.
8. Sra, M.; Garrido-Jurado, S.; Maes, P. Oasis: Procedurally Generated Social Virtual Spaces from 3D Scanned Real Spaces. *IEEE Trans. Vis. Comput. Graph.* **2018**, *24*, 3174–3187. [CrossRef]
9. Zhang, S. High-speed 3D shape measurement with structured light methods: A review. *Opt. Lasers Eng.* **2018**, *106*, 119–131. [CrossRef]
10. Gorthi, S.S.; Rastogi, P. Fringe projection techniques: Whither we are? *Opt. Lasers Eng.* **2010**, *48*, 133–140. [CrossRef]
11. Geng, J. Structured-light 3D surface imaging: A tutorial. *Adv. Opt. Photonics* **2011**, *3*, 128–160. [CrossRef]
12. Zuo, C.; Feng, S.; Huang, L.; Tao, T.; Yin, W.; Chen, Q. Phase shifting algorithms for fringe projection profilometry: A review. *Opt. Lasers Eng.* **2018**, *109*, 23–59. [CrossRef]
13. Zhang, Z. Review of single-shot 3D shape measurement by phase calculation-based fringe projection techniques. *Opt. Lasers Eng.* **2012**, *50*, 1097–1106. [CrossRef]
14. Feng, S.; Zhang, L.; Zuo, C.; Tao, T.; Chen, Q.; Gu, G. High dynamic range 3-D measurements with fringe projection profilometry: A review. *Meas. Sci. Technol.* **2018**, *29*, 122001. [CrossRef]
15. Jiang, H.; Zhao, H.; Li, X. High dynamic range fringe acquisition: A novel 3-D scanning technique for high-reflective surfaces. *Opt. Lasers Eng.* **2012**, *50*, 1484–1493. [CrossRef]
16. Zhang, S.; Yau, S.T. High dynamic range scanning technique. *Opt. Eng.* **2009**, *48*, 033604.
17. Salahieh, B.; Chen, Z.; Rodriguez, J.J.; Liang, R. Multi-polarization fringe projection imaging for high dynamic range objects. *Opt. Express* **2014**, *22*, 10064–10071. [CrossRef]
18. Feng, S.; Zhang, Y.; Chen, Q.; Zuo, C.; Li, R.; Shen, G. General solution for high dynamic range three-dimensional shape measurement using the fringe projection technique. *Opt. Lasers Eng.* **2014**, *59*, 56–71. [CrossRef]
19. Liu, G.H.; Liu, X.Y.; Feng, Q.Y. 3D shape measurement of objects with high dynamic range of surface reflectivity. *Appl. Opt.* **2011**, *50*, 4557–4565. [CrossRef]
20. Jiang C.; Bell T.; Zhang S. High dynamic range real-time 3D shape measurement. *Opt. Express* **2016**, *24*, 7337–7346. [CrossRef]
21. Lin, H.; Gao, J.; Mei, Q.; He, Y.; Liu, J. Adaptive digital fringe projection technique for high dynamic range three-dimensional shape measurement. *Opt. Express* **2016**, *24*, 7703–7718. [CrossRef]

22. Waddington, C.J.; Kofman, J.D. Modified sinusoidal fringe-pattern projection for variable illuminance in phase-shifting three-dimensional surface-shape metrology. *Opt. Eng.* **2014**, *53*, 084109. [CrossRef]
23. Sheng, H.; Xu, J.; Zhang, S. Dynamic projection theory for fringe projection profilometry. *Appl. Opt.* **2017**, *56*, 8452–8460. [CrossRef]
24. Chen, C.; Gao, N.; Wang, X.; Zhang, Z. Adaptive pixel-to-pixel projection intensity adjustment for measuring a shiny surface using orthogonal color fringe pattern projection. *Meas. Sci. Technol.* **2018**, *29*, 055203. [CrossRef]
25. Qi, Z.; Wang, Z.; Huang, J.; Xue, Q. Improving the quality of stripes in structured-light three-dimensional profile measurement. *Opt. Eng.* **2016**, *56*, 031208. [CrossRef]
26. Haug, K.; Pritschow, G. Robust laser-stripe sensor for automated weld-seam-tracking in the shipbuilding industry. In Proceedings of the 24th Annual Conference of the IEEE Industrial Electronics Society, Aachen, Germany, 31 August–4 September 1998; Volume 2, pp. 1236–1241.
27. Usamentiaga, R.; Molleda, J.; Garcia, D.F.; Bulnes, F.G. Removing vibrations in 3D reconstruction using multiple laser stripes. *Opt. Lasers Eng.* **2014**, *53*, 51–59. [CrossRef]
28. Ye, Z.; Lianpo, W.; Gu, Y.; Chao, Z.; Jiang, B.; Ni, J. A Laser Triangulation-Based 3D Measurement System for Inner Surface of Deep Holes. In Proceedings of the ASME 2018 13th International Manufacturing Science and Engineering Conference, College Station, TX, USA, 18–22 June 2018.
29. Usamentiaga, R.; Molleda, J.; Garcia, D.F. Structured-light sensor using two laser stripes for 3D reconstruction without vibrations. *Sensors* **2014**, *14*, 20041–20063. [CrossRef]
30. Sun, Q.; Chen, J.; Li, C. A robust method to extract a laser stripe centre based on grey level moment. *Opt. Lasers Eng.* **2015**, *67*, 122–127. [CrossRef]
31. Tian, Q.; Zhang, X.; Ma, Q.; Ge, B. Utilizing polygon segmentation technique to extract and optimize light stripe centerline in line-structured laser 3D scanner. *Pattern Recognit.* **2016**, *55*, 100–113.
32. Usamentiaga, R.; Molleda, J.; García, D.F. Fast and robust laser stripe extraction for 3D reconstruction in industrial environments. *Mach. Vis. Appl.* **2012**, *23*, 179–196. [CrossRef]
33. Zhang, S. *High-Speed 3D Imaging with Digital Fringe Projection Techniques*; CRC Press: Boca Raton, FL, USA, 2016.
34. Huang, P.S.; Zhang, C.; Chiang, F.P. High-speed 3-D shape measurement based on digital fringe projection. *Opt. Eng.* **2003**, *42*, 163–169. [CrossRef]
35. Holmstrom, S.T.; Baran, U.; Urey, H. MEMS laser scanners: A review. *J. Microelectromech. Syst.* **2014**, *23*, 259–275. [CrossRef]
36. Wang, H.; Zhou, L.; Zhang, X.; Xie, H. Thermal Reliability Study of an Electrothermal MEMS Mirror. *IEEE Trans. Device Mater. Reliab.* **2018**, *18*, 422–428. [CrossRef]
37. Xie, H. Editorial for the Special Issue on MEMS Mirrors. *Micromachines* **2018**, *9*, 99. [CrossRef]
38. Tanguy, Q.A.; Duan, C.; Wang, W.; Xie, H.; Bargiel, S.; Struk, P.; Lutz, P.; Gorecki, C. A 2-axis electrothermal MEMS micro-scanner with torsional beam. In Proceedings of the 2016 International Conference on Optical MEMS and Nanophotonics (OMN), Singapore, 31 July–4 August 2016; pp. 1–2.
39. Yoshizawa, T.; Wakayama, T. Compact camera system for 3D profile measurement. In Proceedings of the 2009 International Conference on Optical Instruments and Technology: Optoelectronic Imaging and Process Technology, Shanghai, China, 19–22 October 2009; Volume 7513, p. 751304.
40. Guo, H.; He, H.; Chen, M. Gamma correction for digital fringe projection profilometry. *Appl. Opt.* **2004**, *43*, 2906–2914. [CrossRef]
41. Lu, J.; Mo, R.; Sun, H.; Chang, Z. Flexible calibration of phase-to-height conversion in fringe projection profilometry. *Appl. Opt.* **2016**, *55*, 6381–6388. [CrossRef]

Article

FR4-Based Electromagnetic Scanning Micromirror Integrated with Angle Sensor

Hongjie Lei [1,2,*], Quan Wen [1,2,3,*], Fan Yu [1,2], Ying Zhou [1,2] and Zhiyu Wen [1,2]

[1] Microsystem Research Center, College of Optoelectronic Engineering, Chongqing University, Chongqing 400044, China; Yu_Fan@cqu.edu.cn (F.Y.); yzhou@cqu.edu.cn (Y.Z.); wzy@cqu.edu.cn (Z.W.)
[2] Key Laboratory of Fundamental Science of Micro/Nano-Device and System Technology, Chongqing University, Chongqing 400044, China
[3] Fraunhofer Institute for Electronic Nano Systems (ENAS), 09131 Chemnitz, Germany
* Correspondence: leihongjie@cqu.edu.cn (H.L.); Quan.Wen@enas.fraunhofer.de (Q.W.); Tel.: +86-023-6511-1010 (H.L.); +49(0)-371-4500-1252 (Q.W.)

Received: 30 March 2018; Accepted: 26 April 2018; Published: 2 May 2018

Abstract: This paper presents a flame retardant 4 (FR4)-based electromagnetic scanning micromirror, which aims to overcome the limitations of conventional microelectromechanical systems (MEMS) micromirrors for the large-aperture and low-frequency scanning applications. This micromirror is fabricated through a commercial printed circuit board (PCB) technology at a low cost and with a short process cycle, before an aluminum-coated silicon mirror plate with a large aperture is bonded on the FR4 platform to provide a high surface quality. In particular, an electromagnetic angle sensor is integrated to monitor the motion of the micromirror in real time. A prototype has been assembled and tested. The results show that the micromirror can reach the optical scan angle of 11.2° with a low driving voltage of only 425 mV at resonance (361.8 Hz). At the same time, the signal of the integrated angle sensor also shows good signal-to-noise ratio, linearity and sensitivity. Finally, the reliability of the FR4 based micro-mirror has been tested. The prototype successfully passes both shock and vibration tests. Furthermore, the results of the long-term mechanical cycling test (50 million cycles) suggest that the maximum variations of resonant frequency and scan angle are less than 0.3% and 6%, respectively. Therefore, this simple and robust micromirror has great potential in being useful in a number of optical microsystems, especially when large-aperture or low-frequency is required.

Keywords: scanning micromirror; electromagnetic actuator; angle sensor; flame retardant 4 (FR4)

---

1. Introduction

The scanning micromirror is a promising component for wide applications, such as projection displays [1], barcode readers [2], micro-spectrometers [3,4] and biomedical imaging [5]. Currently, most of the scanning micromirrors are developed and fabricated using the microelectromechanical systems (MEMS) technology and can be driven by different actuation mechanisms, such as electrostatic [6,7], electrothermal [3,8], electromagnetic [4,9,10] and piezoelectric mechanisms [11,12]. The silicon MEMS micromirror shows exceptional properties that are suitable for high frequency scanning applications due to its small size, low power consumption and fast speed [13]. However, the low-frequency (in the order of a few hundred Hz) silicon MEMS micromirror is fragile and cannot survive the environmental shocks and vibrations due to the brittleness of silicon [14]. Moreover, the fatigue strength of silicon decreases when the size of MEMS structure increases [15], thus dramatically limiting the aperture of the mirror plate.

Currently, a low-frequency scanning and large-aperture micromirror is required for a broad spectrum of applications, such as micro-spectrometers [4], laser projection [16–18], fluorescence microscopes [19] and so on. It has been a continuous and ongoing task to find a proper alternative in order to meet the vast application demand. Some groups have proposed the use of metal instead of silicon as the substrate material in the scanning micromirror devices [19–22]. The metal-based micromirror possesses stronger robustness due to the ductile properties of the metal compared with silicon. However, the metal substrate usually needs an additional separation process to form actuators, inevitably increasing the process complexity and the production cost. Moreover, the surface quality of the metal-based micromirror is inferior to that of the silicon micromirror. Another inexpensive yet highly suitable candidate is flame retardant4 (FR4) [17,18,23], which is inherently a soft material with a low Young's modulus of about 20 Mpa. It is the most widely used material for printed circuit boards (PCB) due to its good electrical, mechanical and thermal properties. Thus, a robust scanning micromirror can be quickly fabricated using FR4 as the substrate through the commercially available and low-cost PCB fabrication process.

In this paper, a FR4-based electromagnetic scanning micromirror is proposed in order to overcome the limitations of the conventional MEMS micromirrors for large-aperture and low-frequency scanning applications. The copper coils for the actuation are printed on the bottom layer of a thin FR4 platform. An aluminum-coated silicon mirror plate with a large aperture (11.7 mm × 10.3 mm) is bonded on the FR4 platform to provide a high surface quality. Particularly, an electromagnetic angle sensor with double-layer sensing coil is integrated on the same FR4 platform without the requirement of an additional process. The angle sensor can monitor the deflection angle in real time, which is very useful for a micromirror as it forms precise closed-loop control [24,25]. The innovation lies in that both the actuator and angle sensor are simultaneously fabricated on a FR4 board using a low-cost PCB process instead of the expensive Si-based MEMS process. Furthermore, the device is fully packaged and tested to demonstrate its great performance in terms of driving, sensing and reliability. The rest of this paper is organized as follows. The design and theory of the electromagnetic micromirror integrated with the angle sensor is introduced in Section 2. After this, the tests and corresponding results of actuation, sensing, response and reliability are described in Section 3. Finally, a brief conclusion is given in Section 4.

2. Design and Theory

The proposed FR4-based electromagnetic scanning micromirror is shown in Figure 1. It includes a 400-μm-thick FR4 platform, a 500-μm-thick silicon mirror plate on it and a pair of permanent magnets. The layout of the FR4 platform is shown in Figure 1a. Both the outer single-layer driving coil (on the bottom layer) and the inner double-layer sensing coil are simultaneously integrated into this platform. Furthermore, the top and bottom layers for sensing coil are connected with vias. The 12 mm × 12 mm central platform is anchored to the frame by a pair of torsion bars. The length and width of the torsion bars are 11 mm and 1 mm, respectively. Two permanent magnets are assembled in parallel, generating a magnetic field that is mainly parallel to the mirror. When the driving coil is energized, a Lorentz force is generated to exert a net torque about the torsion axis. Consequently, the platform with the mirror plate is actuated to tilt, which is shown in Figure 1b. At the same time, the sensing coil induces an electromotive force in this platform. Therefore, the deflection angle can be monitored in real time. Figure 1c shows the schematic drawing of the assembled scanning micromirror. Both the FR4 platform and permanent magnets are sandwiched between the baseplate and coverplate.

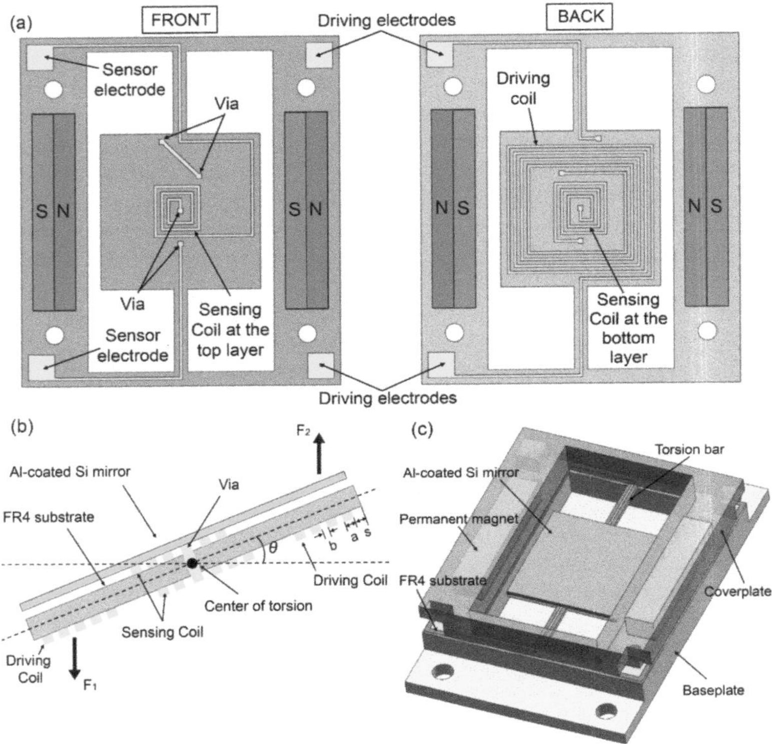

Figure 1. (**a**) Layout of the flame retardant 4 (FR4) platform with double-layer copper coils; (**b**) Electromagnetic actuation and sensing of the FR4 platform with the attached Al-coated Si mirror plate; and (**c**) Schematic drawing of the assembled scanning micromirror.

The motion of the scanning micromirror can be approximated as a forced oscillating system and thus, the corresponding dynamical equation can be expressed as:

$$J_m\ddot{\theta} + C\dot{\theta} + K\theta = T \tag{1}$$

where J_m, C, K, T and θ present the moment of inertia of the platform with the mirror plate around the torsion axis, the damping coefficient, the torsional stiffness of the torsion bars, the net torque and the mechanical half deflection angle, respectively. As the driving signal is sinusoidal with the resonant frequency f, by solving Equation (1), the mechanical half deflection angle can be obtained as:

$$\theta = \frac{T}{K}Q \tag{2}$$

where $Q = 1/2\xi$ is the quality factor and ξ represents the damping ratio.

When the mechanical half deflection angle is small ($\cos\theta \approx 1$), the net torque can be approximately described as:

$$T = iBM_d \tag{3}$$

where

$$\begin{array}{l} M_d = \sum_{n=1}^{N_d} \left(\frac{w-2s-b}{2} - (n-1)(a+b) \right) \times ((2l - 4s - 3a - 5b) - 4(n-1)(a+b)) \\ - \left(\frac{w-2s-b}{2} - (N_d - 1)(a+b) \right) \times ((l - 2s - 2a - 3b) - 2(N_d - 1)(a+b)) \end{array} \quad (4)$$

where i is the driving current; B is the magnetic flux density produced by the magnets; M_d is the total area of all the driving coils; l and w are the length and width of the FR4 platform, respectively; N_d is the number of the driving coil turns; a is the spacing of adjacent coils; b is the width of coil; and s is the width of the exterior border zone of the platform. According to Equations (2)–(4), it is easy to predict the micromirror scan angle.

At the same time, the electromotive force induced in the sensing coil can be approximately described as:

$$\varepsilon_s = BM_s\dot{\theta} \quad (5)$$

where

$$\begin{array}{l} M_s = 2\sum_{n=1}^{N_s} \left((a + \frac{3}{2}b) + (n-1)(a+b)\right) \times ((5a + 7b) + 4(n-1)(a+b)) \\ -\left((a + \frac{3}{2}b) + (N_s - 1)(a+b)\right) \times ((3a + 4b) + 2(N_s - 1)(a+b)) \\ +\left(\frac{w}{2} - s\right) \times \left((a + \frac{3}{2}b) + (N_s - 1)(a+b) + \left(\frac{l}{2} - s\right)\right) \end{array} \quad (6)$$

where M_s is the total area of the all the sensing coils; and N_s is the number of the sensing coil turns. All the aforementioned parameters of the FR4 platform are listed in Table 1. The mechanical half deflection angle θ can be described by means of the maximum mechanical half deflection angle θ_0 and the phase shift φ with respect to the driving voltage as follows:

$$\theta = \theta_0 \sin(2\pi f t + \varphi) \quad (7)$$

By substituting Equation (7) to Equation (5), the electromotive force induced in the sensing coil can be expressed as:

$$\varepsilon_s = 2\pi f B M_s \theta_0 \cos(2\pi f t + \varphi) \quad (8)$$

Therefore, the scanning angle can be attained in real time through the voltage output of the angle sensor. Moreover, the amplitude of the induced electromotive force is proportional to the maximum deflection angle of the micromirror.

Table 1. Parameters of the flame retardant 4 (FR4) platform integrated with driving and sensing coils.

Parameters	l	w	N_d	N_s	a	b	s
Value	12 mm	12 mm	11	9	0.1 mm	0.1 mm	0.5 mm

3. Test Results of Prototype

3.1. Optical Scan Angle

Figure 2 shows the prototype of the FR4-based electromagnetic scanning micromirror with a simple plexiglass package. The performance of the device is tested by shooting a laser spot on the mirror plate and measuring the length of the projected laser line. After this, the scan angle can be calculated according to the length of the projected laser line and the distance between the projected screen and the mirror plate. A sinusoidal voltage is applied to the driving electrodes to actuate the micromirror. Therefore, it is convenient to find the resonant frequency and test the actuation performance by adjusting the driving signal frequency and driving voltage amplitude, respectively. Figure 3a shows the results of the micromirror optical scan angle with changes in the frequency. According to the test results, its resonant frequency is 361.8 Hz when it reaches the maximum scan angle and the 3-dB bandwidth is 6.16 Hz. Hence, the quality factor of $Q = 59$ and damping ratio of $\xi = 0.0085$ can be obtained by the half-power bandwidth method. Figure 3b shows the relationship between the optical scan angle and the driving voltage amplitude when the driving frequency is fixed at its resonant frequency of 361.8 Hz. The optical scan angle increases with an increase in the driving voltage and can reach the maximum value of 11.2° at 425 mV. This result is very close to the theoretical value. The slightly non-linear nature of the test curve could be caused by the non-linear spring effect of the torsion bars [26]. The measured resistances of the driving coil and sensing coil are 3.7 Ω and 2.4 Ω, respectively.

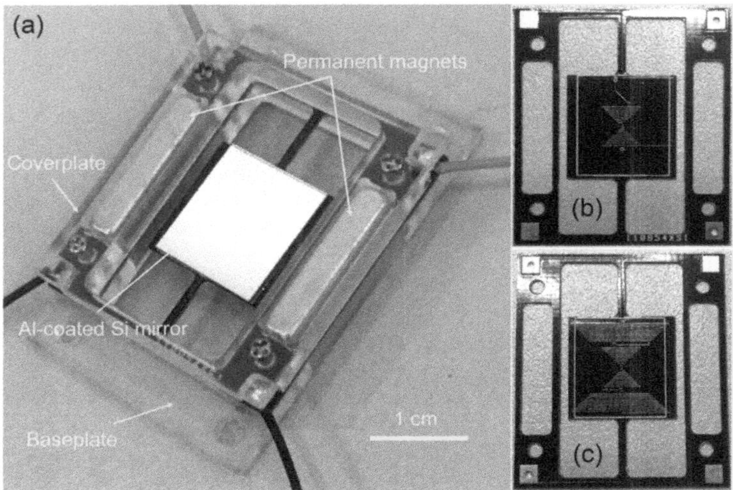

Figure 2. (**a**) The photograph of the prototype of FR4-based electromagnetic scanning micromirror with a simple plexiglass package; (**b**) Front-side of the FR4 platform integrated with copper coils for sensing; and (**c**) Back-side of the FR4 platform integrated with copper coils for driving and sensing.

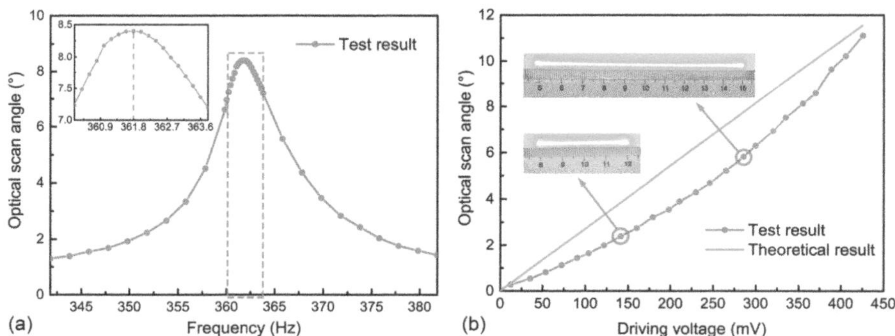

Figure 3. Optical scan angle test of the FR4-based scanning micromirror: (**a**) Optical scan angle versus frequency with fixed driving voltage and the resistance of the driving coil being 3.7 Ω; and (**b**) Optical scan angle versus driving voltage at its resonant frequency.

3.2. Angle Sensor

The output signal of the angle sensor is amplified by a simple amplifier with a gain of 400. After this, it can be measured using an oscilloscope. As seen in Figure 4a, the sensor output shows a good signal-to-noise ratio and is approximately in phase with the driving signal. This result indicates the good feasibility of using the sensing coil as an angle sensor. The value of the signal-to-noise ratio can be obtained as 43 dB through a fast Fourier transform for the sensor output in the time domain. Furthermore, the phase relation can be predicted theoretically. According to Equation (5), the electromotive force induced in the sensing coil is in phase with the angular velocity of the micromirror, while the angular velocity is in phase with the driving voltage when the micromirror is actuated at its resonant frequency. Consequently, the electromotive force is in phase with the driving voltage, which means that the resonance can be tracked by monitoring the phase difference between the driving and sensing signals.

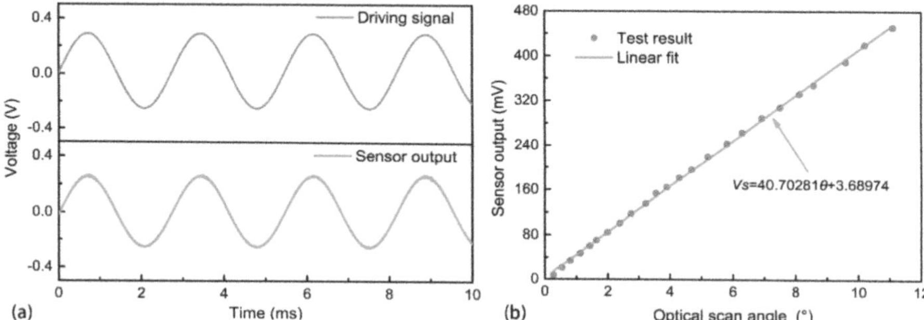

Figure 4. Measurement of the integrated angle sensor: (**a**) Time dependence of the driving signal and the sensor output signal; and (**b**) Relation between the sensor output signal and the optical scan angle.

Figure 4b plots the relationship between the sensor output voltage and the micromirror optical scan angle at the resonance point. The sensor signal is proportional to the optical scan angle of the micromirror. We determined a significant linear relation with a correlation coefficient of $r = 0.99963$. This is consistent with the theoretical relation determined by Equation (8). The sensitivity of the angle sensor can be determined as 40.70281 mV/°, which is enough for feedback control [12]. According to Equation (8), the theoretical value of sensitivity is 44.2131 mV/°. The difference between the theoretical and experimental values could be attributed to the fabrication error and coupling influence between the driving and sensing coils.

3.3. Response Test

Due to the integration of the angle sensor, it is very convenient to test the dynamic response of the micromirror through the sensor signal without the need of an external experimental setup. A sinusoidal driving signal of 361.8 Hz, which is the resonant frequency, is applied. The results are shown in Figure 5. According to the test result in Figure 5a, it takes about 120 ms for the micromirror to reach stable oscillation. This short time will not cause any noticeable unstable scanning delay. Furthermore, the damping ratio can be obtained by the free vibration decay method shown in Figure 5b, with the corresponding value being 0.0081. The result obtained from the angle sensor is very close to the aforementioned value obtained from the micromirror scanning. This small error can be attributed to the measuring error and the difference between the two test methods of the damping ratio. Therefore, the accuracy of the sensor signal is further verified by the test result.

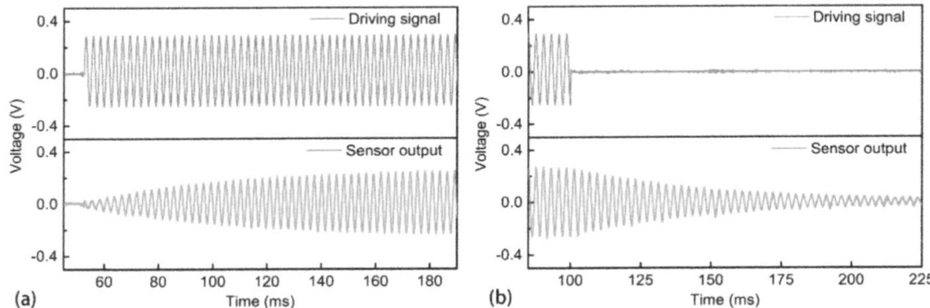

Figure 5. Test of: (**a**) the transient response and (**b**) damping ratio of the micromirror through the integrated angle sensor.

3.4. Reliability of Micromirror

The mechanical shock, vibration, and long-term cycling tests are all performed to evaluate the reliability of FR4-based electromagnetic scanning micromirror. In the mechanical shock, the prototype is tested in three mutually perpendicular axes. The amplitude of the shock is 1000 g with a duration of 1 ms. The results show that the prototype passes the shock test successfully.

After this, the vibration test is carried out by mounting the prototype on the vibration machine. The amplitude of the vibration is 20 g with the sweeping frequency from 20 Hz to 1000 Hz and back in 20 min. It also passes the vibration test without any failure in the three mutually perpendicular axes.

Finally, we perform a long-term cycling test on the prototype by keeping it in resonance for nearly 50 million cycles. Figure 6 plots the variation of the resonant frequency and scan angle. The maximum variations of both parameters are less than 0.3% and 6%, respectively, which demonstrates its great reliability and resistance against the stress and heat of FR4. The minor changes can be explained by the changing temperature and humidity in the laboratory environment during the nearly 40-hour test period [19]. Furthermore, the angle variation can be reduced or even eliminated by using the integrated angle sensor to form a precise close-loop feedback control system.

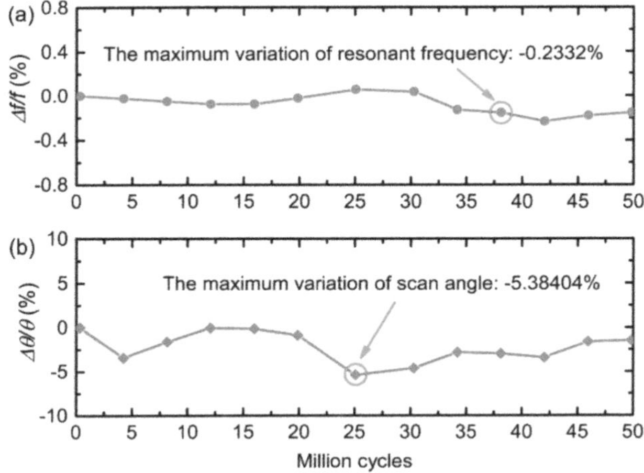

Figure 6. Long-term cycling test on the prototype by keeping it in resonance for nearly 50 million cycles, which shows the variation of: (**a**) resonant frequency and (**b**) scan angle in the test.

4. Conclusions

In summary, a prototype of FR4-based electromagnetic scanning micromirror integrated with an angle sensor is presented in this paper. It is designed and fabricated based on a commercial PCB technology for a short time period at a low cost. The device has a large aperture (11.7 mm × 10.3 mm) with high surface quality and low frequency, which is difficult to achieve with the conventional silicon MEMS. The test results show that the optical scan angle can reach 11.2° under a low driving voltage of only 425 mV at the resonant frequency of 361.8 Hz. Additionally, the signal of the integrated angle sensor is confirmed to be proportional to the optical scan angle of the micromirror. It has good signal-to-noise ratio, linearity and sensitivity. Thus, it can be used for the real-time angle monitoring and precise close-loop control. Finally, the reliability tests, including shock, vibration and long-term cycling, are carried out. The results show that the prototype possesses high reliability. Therefore, this simple and robust micromirror holds promise for a number of optical microsystem applications that require low-frequency scanning. In the future, the performance of the actuation and sensing can be further improved by employing a multilayer PCB process to increase the number of coil turns and some simple electronics can be even integrated on the same FR4 platform.

Author Contributions: H.L., Q.W. and Z.W. conceived and designed the device; H.L., Q.W. and Y.Z. deduced the theory. H.L. and F.Y. performed the test of prototype and analyzed the data; H.L. and Q.W. wrote and revised the paper respectively.

Acknowledgments: This work is supported by the Special Funds of the National Natural Science Foundation of China (Grant No. 61327002) and the Fundamental Research Funds for Central Universities (Grant No. 106112017CDJXY120006).

Conflicts of Interest: The authors declare no conflict of interest.

References

1. Baran, U.; Brown, D.; Holmstrom, S.; Balma, D.; Davis, W.O.; Muralt, P.; Urey, H. Resonant PZT MEMS scanner for high-resolution displays. *J. Microelectromech. Syst.* **2012**, *21*, 1303–1310. [CrossRef]
2. Yalcinkaya, A.D.; Ergeneman, O.; Urey, H. Polymer magnetic scanners for bar code applications. *Sens. Actuators A Phys.* **2007**, *135*, 236–243. [CrossRef]
3. Han, F.; Wang, W.; Zhang, X.; Xie, H. Modeling and control of a large-stroke electrothermal MEMS mirror for fourier transform microspectrometers. *J. Microelectromech. Syst.* **2016**, *25*, 750–760. [CrossRef]
4. Zhou, Y.; Wen, Q.; Wen, Z.; Huang, J.; Chang, F. An electromagnetic scanning mirror integrated with blazed grating and angle sensor for near infrared micro spectrometer. *J. Micromech. Microeng.* **2017**, *27*, 125009. [CrossRef]
5. Tang, S.; Jung, W.; Mccormick, D.; Xie, T.; Su, J.; Ahn, Y.C.; Tromberg, B.J.; Chen, Z. Design and implementation of fiber-based multiphoton endoscopy with microelectromechanical systems scanning. *J. Biomed. Opt.* **2009**, *14*, 034005. [CrossRef] [PubMed]
6. Hung, C.L.; Lai, Y.H.; Lin, T.W.; Fu, S.G.; Lu, S.C. An electrostatically driven 2D micro-scanning mirror with capacitive sensing for projection display. *Sens. Actuators A Phys.* **2015**, *222*, 122–129. [CrossRef]
7. Moon, S.; Lee, J.; Yun, J.; Lim, J. Two-axis electrostatic gimbaled mirror scanner with self-aligned tilted stationary combs. *IEEE Photonics Technol. Lett.* **2015**, *28*, 557–560. [CrossRef]
8. Tanguy, Q.; Bargiel, S.; Xie, H.; Passilly, N.; Barthès, M.; Gaiffe, O.; Rutkowski, J.; Lutz, P.; Gorecki, C.; Tanguy, Q. Design and fabrication of a 2-axis electrothermal MEMS micro-scanner for optical coherence tomography. *Micromachines* **2017**, *8*, 146. [CrossRef]
9. Han, A.; Cho, A.R.; Ju, S.; Ahn, S.H.; Bu, J.U.; Ji, C.H. Electromagnetic biaxial vector scanner using radial magnetic field. *Opt. Express* **2016**, *24*, 15813. [CrossRef] [PubMed]
10. Miyajima, H.; Asaoka, N.; Isokawa, T.; Ogata, M.; Aoki, Y.; Imai, M.; Fujimori, O.; Katashiro, M.; Matsumoto, K. A MEMS electromagnetic optical scanner for a commercial confocal laser scanning microscope. *J. Microelectromech. Syst.* **2003**, *12*, 243–251. [CrossRef]
11. Koh, K.H.; Lee, C. A two-dimensional MEMS scanning mirror using hybrid actuation mechanisms with low operation voltage. *J. Microelectromech. Syst.* **2012**, *21*, 1124–1135. [CrossRef]

12. Naono, T.; Fujii, T.; Esashi, M.; Tanaka, S. A large-scan-angle piezoelectric MEMS optical scanner actuated by a Nb-doped PZT thin film. *J. Micromech. Microeng.* **2014**, *24*, 5010. [CrossRef]
13. Holmstrom, S.T.S.; Baran, U.; Urey, H. MEMS laser scanners: A review. *J. Microelectromech. Syst.* **2014**, *23*, 259–275. [CrossRef]
14. Ataman, Ç.; Urey, H. Compact fourier transform spectrometers using FR4 platform. *Sens. Actuators A Phys.* **2009**, *151*, 9–16. [CrossRef]
15. Namazu, T.; Isono, Y. Fatigue life prediction criterion for micro–nanoscale single-crystal silicon structures. *J. Microelectromech. Syst.* **2009**, *18*, 129–137. [CrossRef]
16. Jeong, H.M.; Park, Y.H.; Lee, J.H. Slow scanning electromagnetic scanner for laser display. *J. Microlithogr. Microfabr. Microsyst.* **2008**, *7*, 1589–1604.
17. Zuo, H.; He, S. FPCB micromirror-based laser projection availability indicator. *IEEE Trans. Ind. Electron.* **2016**, *63*, 3009–3018. [CrossRef]
18. Zuo, H.; He, S. FPCB ring-square electrode sandwiched micromirror based laser pattern pointer. *IEEE Trans. Ind. Electron.* **2017**, *64*, 6319–6329. [CrossRef]
19. Wang, Y.; Gokdel, Y.D.; Triesault, N.; Wang, L.; Huang, Y.Y.; Zhang, X. Magnetic-Actuated Stainless Steel Scanner for Two-Photon Hyperspectral Fluorescence Microscope. *J. Microelectromech. Syst.* **2014**, *23*, 1208–1218. [CrossRef]
20. Park, J.H.; Akedo, J.; Sato, H. High-speed metal-based optical microscanners using stainless-steel substrate and piezoelectric thick films prepared by aerosol deposition method. *Sens. Actuators A Phys.* **2007**, *135*, 86–91. [CrossRef]
21. Gokdel, Y.D.; Sarioglu, B.; Mutlu, S.; Yalcinkaya, A.D. Design and fabrication of two-axis micromachined steel scanners. *J. Micromech. Microeng.* **2009**, *19*, 1403–1407. [CrossRef]
22. Ye, L.; Zhang, G.; You, Z. Large-aperture kHz operating frequency Ti-alloy based optical micro scanning mirror for LiDAR application. *Micromachines* **2017**, *8*, 120. [CrossRef]
23. Urey, H.; Holmstrom, S.; Yalcinkaya, A.D. Electromagnetically actuated FR4 scanners. *IEEE Photonics Technol. Lett.* **2008**, *20*, 30–32. [CrossRef]
24. Tseng, V.F.G.; Xie, H. Simultaneous piston position and tilt angle sensing for large vertical displacement micromirrors by frequency detection inductive sensing. *Appl. Phys. Lett.* **2015**, *107*, 214102. [CrossRef]
25. Kobayashi, T.; Maeda, R.; Itoh, T.; Sawada, R. Smart optical microscanner with piezoelectric resonator, sensor, and tuner using Pb(Zr, Ti)O_3 thin film. *Appl. Phys. Lett.* **2007**, *90*, 183514. [CrossRef]
26. Kundu, S.K.; Ogawa, S.; Kumagai, S.; Fujishima, M.; Hane, K.; Sasaki, M. Nonlinear spring effect of tense thin-film torsion bar combined with electrostatic driving. *Sens. Actuators A Phys.* **2013**, *195*, 83–89. [CrossRef]

MDPI

Article

A MEMS Variable Optical Attenuator with Ultra-Low Wavelength-Dependent Loss and Polarization-Dependent Loss

Huangqingbo Sun [1], Wei Zhou [2], Zijing Zhang [1] and Zhujun Wan [1,3,*]

1 School of Optical and Electronic Information, Huazhong University of Science and Technology, Wuhan 430074, China; u201514056@hust.edu.cn (H.S.); zijing_zhang@hust.edu.cn (Z.Z.)
2 AOFSS (Shenzhen) Co., Ltd., Shenzhen 518103, China; david.zhou@aofss.com
3 Shenzhen Huazhong University of Science and Technology Research Institute, Shenzhen 518000, China
* Correspondence: zhujun.wan@hust.edu.cn; Tel.: +86-27-8755-6188(3092)

Received: 16 October 2018; Accepted: 27 November 2018; Published: 29 November 2018

Abstract: Applications in broadband optical fiber communication system need variable optical attenuators (VOAs) with low wavelength-dependent loss (WDL). Based on analysis on the dispersion of the optical system of a MEMS-based VOA, we provide a method to reduce the WDL significantly with minor revision on the end-face angle of the collimating lens. Two samples are assembled, and the measured WDL is <0.4 dB over the C-band (1.53–1.57 μm) at a 0–20 dB attenuation range. Meanwhile, the new structure helps to reduce the polarization-dependent loss (PDL) to <0.15 dB, which is only half that of conventional devices.

Keywords: variable optical attenuator (VOA); wavelength dependent loss (WDL); polarization dependent loss (PDL); micro-electro-mechanical systems (MEMS)

---

1. Introduction

A variable optical attenuator (VOA) is an important optical device for optical fiber communication and optical instrumentation [1,2]. The main approaches for a VOA include: thermo-optically adjusted Mach–Zehnder interferometer (MZI) based on a planar lightwave circuit (PLC) [3], optical fluid driven by a pump [4,5], liquid-core fiber driven by thermo-optical effect [6], and MEMS technology. Among the variable technologies, MEMS technology is one of the most favorable approaches for a VOA. Chengkuo Lee's group did much work to develop different mechanisms for MEMS VOAs, such as retro-reflective mirrors driven by electro-thermal actuators [7], and reflective mirrors driven by rotary comb drive actuators [8]. The most applicable mechanisms are the MEMS shutter [9,10] and the MEMS torsion mirror [11].

One application of VOAs is in erbium-doped fiber amplifier (EDFA) modules for optical fiber communication. An EDFA usually amplifies broadband optical signals over 1.53–1.57 μm. The same amplification for different wavelengths is required. A VOA is employed in the EDFA module to control the optical power dynamically, which requires it to generate nearly the same attenuation for different wavelengths over the bandwidth of 40 nm. However, the existence of wavelength-dependent loss (WDL) means that the VOA generates different attenuation for different wavelengths. The MEMS torsion mirror, driven by comb drive actuators is characterized by low power consumption and ease of packaging, which enable a VOA to be created with low voltage, small size, and low cost. Thus, VOAs based on a MEMS torsion mirror are widely employed in optical fiber communication. However, they are confounded by high WDLs, especially when operating at a high attenuation level. The conventional VOAs without optimization usually show WDLs of more than 1 dB. Reducing the WDL helps to improve the specifications of the EDFA modules.

Researchers from Mega Sense Inc. presented an optimization method by axially rotating the lens with respect to the fibers, while the process is rather time-consuming [12]. Researchers from JDS Uniphase Corp. optimized the WDL by introducing a wedge prism, which added to the complexity of the device [13]. Researchers from NeoPhotonics Corp. analyzed the cause of WDL extensively, and provided an optimization method, while it requires the collimating lens to be fabricated with high dispersion glass [14].

For applications such as optical instrumentation, PDL is another important parameter for VOAs. The existence of PDL means that the VOA generates different attenuations for different polarizations of light. Reducing PDL helps to improve the precision of the optical instruments. PDL is usually introduced by angular surfaces (which are necessary for reduction of back-reflection) and stress in the optical system. The proposal in [14] requires that the angle of the collimating lens is more than 10 degrees, which will introduce more PDL.

Based on the work in [13,14], this paper systematically addressed the WDL and PDL problems in a VOA based on a MEMS torsion mirror. A simple solution for both WDL and PDL optimization was presented with no excess optical element and special glass material required.

2. Theories

2.1. Structure of the VOA

The addressed VOA comprises a dual-fiber collimator and a MEMS torsion mirror coaxially assembled, as shown in Figure 1. The collimator includes a plano-convex collimating lens and two single-mode fibers (SMFs) fixed by a glass capillary. Both the collimating lens and the glass capillary are housed by a glass tube. The fiber facet and the MEMS torsion mirror are located at the front and rear focal planes of the collimating lens, respectively, as shown in Figure 1a,b. The optical signal is input from one optical fiber, and then collimated. The collimated beam is reflected by the mirror and then refocused onto the facet of another optical fiber. The refocused beam spot deviates from the output fiber core, due to the deflection of the mirror, and thus results in the desired attenuation of optical power. In order to reduce back reflection, the facet of the fibers and the planar facet of the lens are both angularly polished. For a conventional structure, both facets are usually polished with an angle of 8° and aligned parallel, as shown in Figure 1c. Based on analysis on the dispersion of the optical system, we provide a new design to optimize the WDL of the VOA. The facet of the fibers keeps unchanged, while the facet of the lens is polished with an angle of −7°. The two facets are aligned as Figure 1d. Meanwhile, the length of the lens is also adjusted, and the theories are shown in Section 2.2.

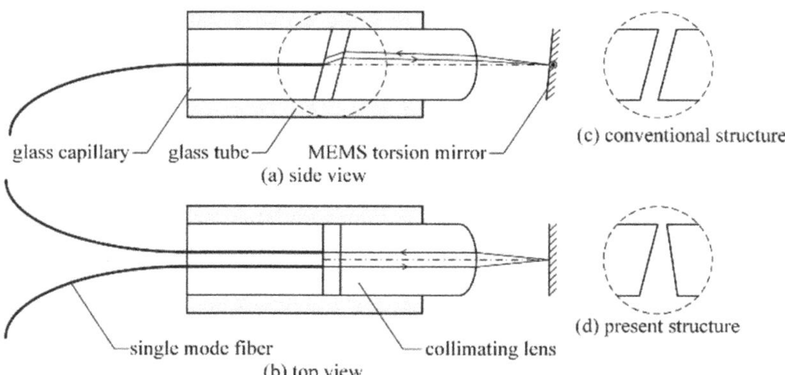

Figure 1. Structure of the variable optical attenuator (VOA) based on a MEMS torsion mirror.

The optical attenuation is tunable by adjusting the tilt angle of the mirror, which results in a lateral offset X of the refocused beam spot on the facet of the output fiber. The attenuation due to lateral offset can be described as [13,15]:

$$A = 4.34\left(\frac{X}{\omega}\right)^2 \tag{1}$$

where ω is the mode radius of the SMF.

2.2. WDL Analysis and Optimization

The mode size $\omega(\lambda)$ of the SMF is wavelength-dependent, and thus results in a wavelength-dependent attenuation $A(\lambda)$ (which is usually called WDL) according to Equation (1). The wavelength dependence of the mode size is rather complicated. However, within a relatively narrow range, such as C-band, the mode size can be linearly approximated. Thus we can assume the mode radii as $\omega_s = \omega_c - \Delta\omega$ and $\omega_l = \omega_c + \Delta\omega$ for the shortest wavelength λ_s and longest wavelength λ_l, respectively, given ω_c as the mode radius of the central wavelength λ_c. According to Equation (1), the WDL before optimization can be obtained as:

$$\mathrm{WDL} = 4.34X^2\left(\frac{1}{\omega_s^2} - \frac{1}{\omega_l^2}\right) = A_c\omega_c^2\left(\frac{1}{\omega_s^2} - \frac{1}{\omega_l^2}\right) \tag{2}$$

which is the attenuation difference between λ_l and λ_s in the C-band. A_c is the attenuation for wavelength λ_c. Equation (2) shows that the WDL adds up when the attenuation level increases. For a VOA operating in the C-band (1.53–1.57 μm), the WDL is usually more than 1 dB at 20 dB attenuation.

According to Equation (1), if we can introduce a wavelength-dependent variable X, then it is possible to reduce the WDL. Considering the dispersion of X, the WDL can be rewritten as:

$$\mathrm{WDL} = 4.34\left[\left(\frac{X_c + \Delta X}{\omega_c + \Delta\omega}\right)^2 - \left(\frac{X_c - \Delta X}{\omega_c - \Delta\omega}\right)^2\right] \tag{3}$$

We assume a linear dispersion for X, with X_c, $X_c + \Delta X$ and $X_c - \Delta X$ as the lateral offset of the λ_c, λ_l and λ_s beam spots, respectively.

According to Equation (3), if Equation (4) is satisfied, the WDL can be reduced to zero for a specified X_c, which corresponds to a certain attenuation level A_c. As shown in Figure 7 of [14], when a specific A_c is selected for zero WDL design, the maximum WDL over the attenuation range (such as 0–20 dB) can be minimized. Here we choose A_c = 13.3 dB for optimization design.

$$\frac{\Delta X}{X_c} = \frac{\Delta\omega}{\omega_c} \tag{4}$$

The dispersion ΔX of the lateral offset results from the dispersion of the optical system. The optical model of the VOA in side view is shown in Figure 2. n_f and n_c are the refractive indexes of the fiber core and the collimating lens, respectively. The focal length of the lens is $f_c = R/(n_c - 1)$, where R is the curvature radius of the right surface. The fiber facet locates at the front focal plane of the collimating lens, and thus, the gap is obtained as $d = f_c - L/n_c$, where L is the length of the lens. The polished angles of the optical fiber and the collimating lens are α and φ, respectively. The deflection of the mirror results in offset X of the beam spot, focused on the output fiber facet. Considering the dispersion effect, the subscript c in n_c and f_c is the parameter corresponding to the central wavelength λ_c.

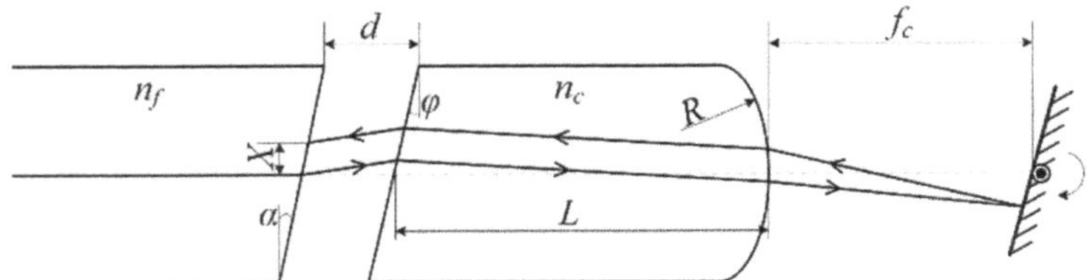

Figure 2. Optical model of the VOA in side view.

Based on the above optical model, the lateral offset X and the dispersion ΔX of the optical system are obtained as Equations (5) and (6) after ray tracing and paraxial approximation:

$$X = X_c + 2\Delta n\left[d\varphi + \frac{d\beta}{n_c - 1} + \frac{\beta L - (n_c - 1)\varphi L}{n_c^2(n_c - 1)}\right] \tag{5}$$

$$\Delta X = 2\Delta n\left[d\varphi + \frac{d\beta}{n_c - 1} + \frac{\beta L - (n_c - 1)\varphi L}{n_c^2(n_c - 1)}\right] \tag{6}$$

where $\beta = (n_f - 1)\alpha$, and $\Delta n = ns - nc = nc - nl$ (the subscripts s and l correspond to the shortest and longest wavelengths in the C-band, respectively) is the difference on refractive index of the lens. By substituting Equation (6) into Equation (4), we obtain Equation (7) as follows,

$$d\varphi + \frac{d\beta}{n_c - 1} + \frac{\beta L - (n_c - 1)\varphi L}{n_c^2(n_c - 1)} = \frac{\Delta\omega X_c}{2\Delta n\omega_c} \tag{7}$$

The fiber employed is SMF-28 by Corning Corp. (Corning, NY, USA) and the glass for the collimating lens is N-SF11 by Schott Corp. (Mainz, Germany). The given parameters are $n_f = 1.4682$, $\Delta\omega = 0.0622$ μm, $\alpha = 8°$ and $n_c = 1.7434$, $\Delta n = 0.00036$, $R = 1.419$ mm. Thus there are only two parameters related by Equation (7), i.e., the facet angle φ and the length L of the lens.

Equation (7) is rather complicated. The correlation between φ and L is numerically plotted, as shown in Figure 3. The lens length is $L = n_c(f_c - d) < n_c f_c = 3.33$ mm. Thus, we just plot the curve with the range of L being 2.6–3.0 mm. Parameters corresponding to any point on the curve can be employed for WDL optimization.

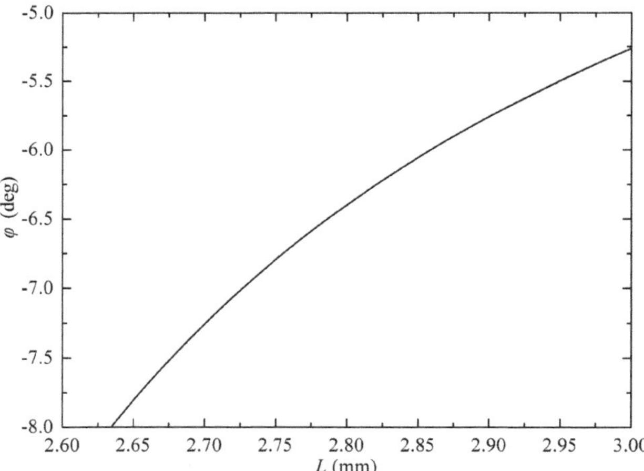

Figure 3. Correlation between the parameters of the collimating lens.

2.3. Return Loss Consideration

In order to reduce the back reflection, the facets of the optical fibers and the collimating lens are usually angled, polished, and aligned, as in Figure 1c. However, we find in Figure 3 that the angle φ must be negative, which means that the facets of the optical fibers and the collimating lens are aligned as in Figure 1d. Thus, the reflected light from the left facet of the lens is more likely to return to the optical fiber than the conventional structure, as shown in Figure 4.

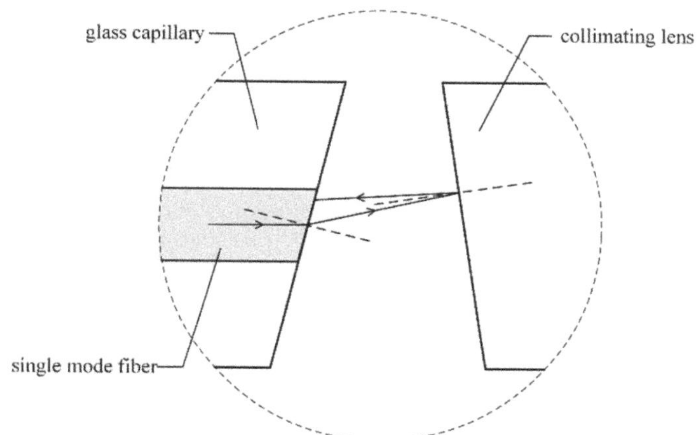

Figure 4. Back flection from the left facet of the collimating lens.

The return loss (RL) can be obtained based on fiber-to-fiber coupling with angular and longitudinal misalignments. By simplifying Equation (33) in [15] with zero lateral misalignment, the RL is obtained as Equation (8), with consideration of the angular misalignment θ and the longitudinal misalignment Z_0, as well as the residual reflection R_r at the AR (anti-reflection)-coated left facet of the lens:

$$RL = -10\log(R_r) - 10\log\left\{\frac{4}{C^2+4}\exp\left[-\frac{kZ_0(C^2+2)\sin^2\theta}{C(C^2+4)}\right]\right\} \tag{8}$$

where $C = \lambda_c Z_0/\pi\omega_c{}^2$. Based on ray tracing and paraxial approximation, the angular and longitudinal misalignments are obtained as $\theta = -2(\varphi + \beta)$ and $Z_0 = 2d$ $(d = f_c - L/n_c)$, respectively. Note that all of the surfaces in the optical system are AR-coated. The reflection from the other facets are negligible.

According to Equation (8), the RL is improved when $|\varphi|$ increases and L decreases. We choose a point from the curve in Figure 3. The corresponding parameters are $\varphi = -7°$ and $L = 2.72$ mm. The residual reflection from the lens facet is $R_r = 0.1\%$ (by sample measurement). Thus, RL is calculated as 53 dB, according to Equation (8).

2.4. PDL Analysis

PDL is usually introduced by angular surfaces and stress in the optical system. The influence of stress depends on the materials and the assembly process, which is outside the scope of this paper. We focus on the influence of the angled surfaces.

Because all of the surfaces in the optical system are AR-coated, the difference between the p-ray and s-ray transmittance introduced by the angular surfaces is negligible. The angular surfaces expand or compress the optical beam only in the tangential plane, and thus result in a slightly elliptical beam spot focused on the facet of the output fiber. When the elliptical beam is received by the circular optical fiber, PDL is introduced.

A circular beam emits from the input fiber, and finally an elliptical beam is focused on the facet of the output fiber. The transformation can be obtained by tracing of the Gaussian beam. Gaussian beam tracing is based on q-parameter and ABCD matrices of the optical elements. When a Gaussian beam is refracted by an angular facet tilted in tangential plane, the ABCD matrix shows different forms in tangential and sagittal planes as Equations (9) and (10), respectively [16].

$$M_T = \begin{pmatrix} \dfrac{\cos\theta_2}{\cos\theta_1} & 0 \\ 0 & \dfrac{n_1\cos\theta_1}{n_2\cos\theta_2} \end{pmatrix} \tag{9}$$

$$M_S = \begin{pmatrix} 1 & 0 \\ 0 & \dfrac{n_1}{n_2} \end{pmatrix} \tag{10}$$

where θ_1 and θ_2 are the incidence and refraction angles, respectively, and n_1 and n_2 are the refractive indices of the materials before and after the angular surface, respectively. Based on Gaussian beam tracing through the entire optical system, the radii of the beam focused on the output fiber facet is obtained. For the conventional structure, the sizes are ω_T = 5.184 μm and ω_S = 5.224 μm (note that the subscripts T and S correspond to the tangential and sagittal planes, respectively), with a difference of $\Delta\omega_{TS}$ = 0.04 μm. For the present structure, the sizes are ω_T = 5.184 μm and ω_S = 5.187 μm, with a difference of $\Delta\omega_{TS}$ = 0.003 μm. The ellipticity is reduced to 1/13, and thus the PDL can be optimized.

3. Experimental Results

Based on above analysis, we designed a MEMS VOA operating at 1.53–1.57 μm. The type of the input/output optical fibers is SMF-28 by Corning Inc. (Corning, NY, USA), with related parameters summarized in Table 1. We keep the facet angle of the optical fiber as 8°, which is the same as the conventional structures. The collimating lens is designed according to the optimization curve in Figure 3. The lens material is N-SF11 by Schott Inc. (Mainz, Germany), with parameters summarized in Table 2. The MEMS mirror is provided by Preciseley Microtechnology Corp. (Edmonton, AB, Canada), and the parameters are summarized in Table 3.

Table 1. Parameters of the optical fiber.

Optical Fiber	Refractive Index	Mode Radius	Mode Dispersion	Facet Angle
Corning SMF-28	n_f = 1.4682	ω_c = 5.2 μm	$\Delta\omega$ = 0.0622 μm	α = 8°

Table 2. Parameters of the collimating lens.

Material	Refractive Index	Length	Curvature Radius	Facet Angle
Schott N-SF11	n_c = 1.7434	L = 2.72 mm	R = 1.419 mm	φ = −7°

Table 3. Parameters of the MEMS mirror.

MEMS Mirror	Clear Aperture	Maximum Tilting Angle
Preciseley LV-VOA	Φ0.85 mm	0.35° @5 V

Figure 5 shows the schematic diagram of the assembly procedures. The MEMS chip was first mounted on a Transister Outline (TO) base, as shown in Figure 6a. Then, a specially designed TO cap was added, with the collimating lens mounted in the cap. Finally, the sub-assembly of MEMS in TO with the lens was aligned with a dual-fiber pigtail by mechanical stages, as shown in Figure 6b. The two parts were fixed with adhesive after optical alignment. According to the above design, the facets of the lens and fiber were aligned as Figure 1d. In real assembly, the enlarged view between the lens and fiber facets is shown in Figure 6c. The final assembly of the MEMS VOA is shown in Figure 7, with the MEMS and optical parts protected by a metal housing and the pigtail fibers protected by plastic tubes of Φ0.9 mm. The samples were assembled in AOFSS (Shenzhen) Ltd., Shenzhen, China, and the product type is LMA-1 B20LV1L1000.

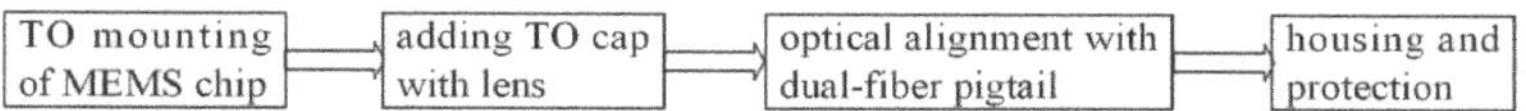

Figure 5. Schematic diagram of the assembling procedures.

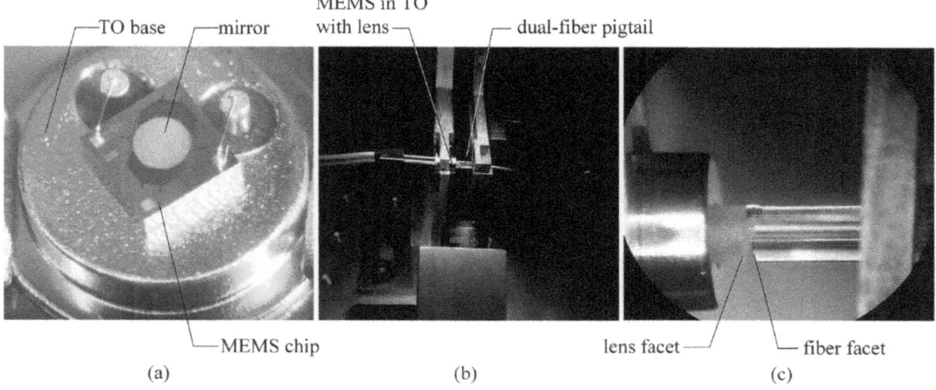

Figure 6. Alignment and assembly of the MEMS VOA. (**a**) MEMS chip assembled on a TO base, (**b**) optical alignment of the VOA, (**c**) enlarged view between the lens and fiber facets.

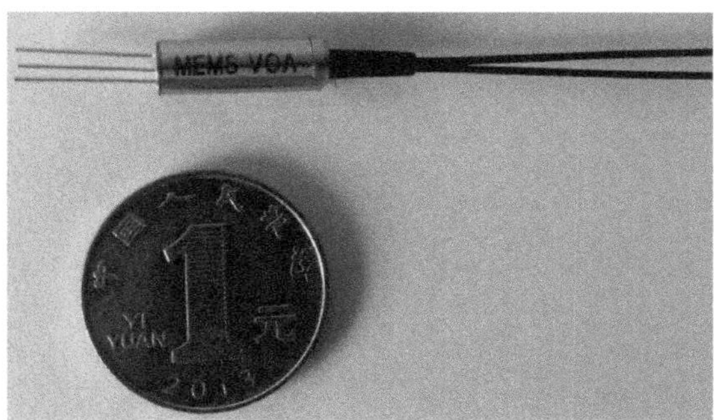

Figure 7. Photograph of the VOA sample.

The measured specifications of two samples are summarized in Table 4, including the insertion loss (IL), *RL*, and PDL at different attenuation levels. For comparison, a conventional MEMS VOA assembled as in Figure 1c was also measured, and the specifications were also summarized in Table 4. As we can see, the new samples show specifications of IL < 0.6 dB, *RL* > 50 dB, and PDL < 0.15 dB over the attenuation range of 0–20 dB, which meet the application requirements. The PDL is reduced to nearly half of the conventional VOA. Meantime, the IL is increased by 0.14 dB, and the *RL* is reduced by 4.8 dB.

Table 4. Measured specifications of the new samples and a conventional MEMS VOA.

No.	IL	*RL*	PDL (dB)				
	(dB)	(dB)	@IL	@5 dB	@10 dB	@15 dB	@20 dB
#1	0.55	52.2	0.02	0.05	0.07	0.10	0.10
#2	0.58	51.5	0.01	0.04	0.05	0.07	0.12
Conventional	0.44	56.3	0.04	0.07	0.13	0.19	0.26

In order to obtain the WDL specifications, the spectra of the new samples and the conventional VOA were measured with an optical spectrum analyzer. Figures 8 and 8 show the spectra of sample #1 and the conventional one at the attenuation level of A_c = 20 dB, respectively. For the former one, the WDL over the C-band @20 dB can be read as 0.28 dB from the spectrum curve, while the latter one is read as 1.07 dB.

The measured WDL over the C-band at different attenuation levels is summarized in Figure 9. The two new samples show a maximum WDL of <0.4 dB, while the maximum WDL of the conventional VOA is 1.07 dB at 20 dB attenuation. The new design reduces the WDL significantly.

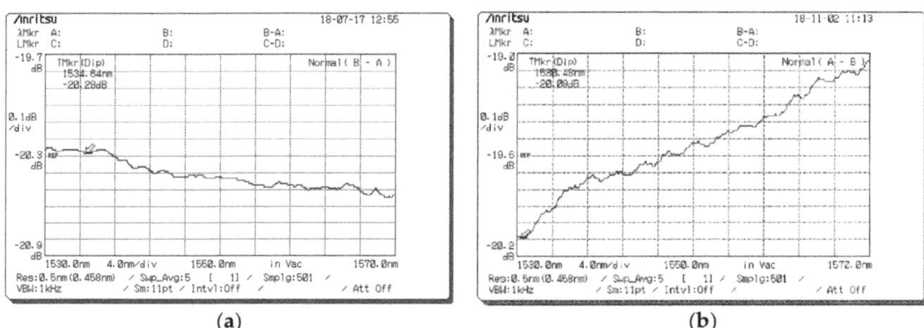

(a) (b)

Figure 8. Measured transmission spectrum at an attenuation level of 20 dB, (**a**) the optimized MEMS VOA, (**b**) the conventional MEMS VOA.

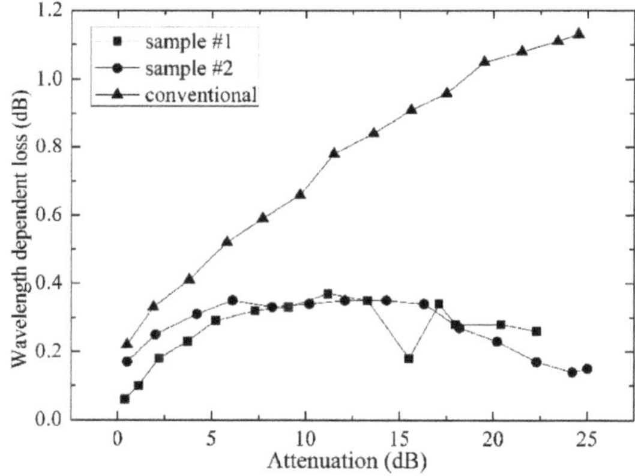

Figure 9. Measured WDL over the C-band at different attenuation levels.

4. Conclusions

We addressed the WDL and PDL problems in a VOA, based on a MEMS torsion mirror. The causes for WDL and PDL were analyzed, and a simple optimization method was presented with the consideration of return loss. The experimental results show that the maximum WDL is <0.4 dB over the wavelength range 1.53–1.57 μm @0–20 dB attenuation range, and the maximum PDL at a 0–20 dB attenuation range is reduced to <0.15 dB, which is only half that of the conventional devices. The results verify the effectiveness of the optimization method well.

Author Contributions: Conceptualization, Z.W.; methodology, H.S. and Z.W.; simulation, H.S.; validation, Z.W., H.S. and Z.Z.; experimentation, W.Z., Z.W. and H.S.; writing, H.S., Z.Z. and Z.W.

Funding: This research was funded by Science, Technology and Innovation Commission of Shenzhen Municipality (JCYJ20170307172638001).

Conflicts of Interest: The authors declare no conflict of interest.

References

1. Tsunashima, S.; Nakajima, F.; Nasu, Y.; Kasahara, R.; Nakanishi, Y.; Saida, T.; Yamada, T.; Sano, K.; Hashimoto, T.; Fukuyama, H.; et al. Silica-based, compact and variable-opticalattenuator integrated coherent receiver with stable optoelectronic coupling system. *Opt. Express* **2012**, *20*, 27174–27179. [CrossRef] [PubMed]
2. Ogawa, I.; Doi, Y.; Hashizume, Y.; Kamei, S.; Tamura, Y.; Ishii, M.; Kominato, T.; Yamazaki, H.; Kaneko, A. Packaging Technology for Ultra-Small Variable Optical Attenuator Multiplexer (V-AWG) With Multichip PLC Integration Structure Using Chip-Scale-Package PD Array. *IEEE J. Sel. Top. Quant.* **2006**, *12*, 1045–1053. [CrossRef]
3. Maese-Novo, A.; Zhang, Z.Y.; Irmscher, G.; Polatynski, A.; Mueller, T.; Felipe, D.D.; Kleinert, M.; Brinker, W.; Zawadzki, C.; Keil, N. Thermally optimized variable optical attenuators on a polymer platform. *Appl. Opt.* **2015**, *54*, 569–575. [CrossRef]
4. Wan, J.; Xue, F.L.; Wu, L.X.; Fu, Y.J.; Hu, J.; Zhang, W.; Hu, F.R. Extensible chip of optofluidic variable optical attenuator. *Opt. Express* **2016**, *24*, 9683–9692. [CrossRef] [PubMed]
5. Duduś, A.; Blue, R.; Uttamchandani, D. Single-mode fiber variable optical attenuator based on a ferrofluid shutter. *App. Opt.* **2015**, *54*, 1952–1957. [CrossRef] [PubMed]
6. Martincek, I.; Pudis, D. Variable Liquid-Core Fiber Optical Attenuator Based on Thermo-Optical Effect. *J. Lightwave Technol.* **2011**, *29*, 2647–2650. [CrossRef]
7. Lee, C. A MEMS VOA Using Electrothermal Actuators. *J. Lightwave Technol.* **2007**, *25*, 490–498. [CrossRef]
8. Andrew, Y.J.; Jiang, S.S.; Lee, C. MOEMS variable optical attenuators using rotary comb drive actuators. *IEEE Photonics Technol. Lett.* **2006**, *18*, 1170–1172.
9. Lee, C. Monolithic-integrated 8CH MEMS variable optical attenuator. *Sens. Actuators A* **2005**, *123*, 596–601. [CrossRef]
10. Unamuno, A.; Blue, R.; Uttamchandani, D. Modeling and Characterization of a Vernier Latching MEMS Variable Optical Attenuator. *J. Microelectromech. Syst.* **2013**, *22*, 1229–1241. [CrossRef]
11. Isamoto, K.; Kato, K.; Morosawa, A.; Chong, C.H.; Fujita, H.; Toshiyoshi, H. A 5-V Operated MEMS Variable Optical Attenuator by SOI Bulk Micromachining. *IEEE J. Sel. Top. Quant.* **2004**, *10*, 570–578. [CrossRef]
12. Chu, C.; Belov, N.; Hout, S.I.; Vanganov, V. Method and Apparatus of Optical Components Having Improved Optical Properties. U.S. Patent 2004/0008967 A1, 15 January 2004.
13. Chen, B.; Liu, X.S.; Yang, Y.T.; Cai, B. Variable Optical Attenuator with Wavelength Dependent Loss Compensation. U.S. Patent 7,295,748 B2, 13 November 2007.
14. Godil, A.A.; Honer, K.; Lawrence, M.; Gustafson, E. Optical Attenuator. U.S. Patent 8,280,218 B2, 2 October 2012.
15. Yuan, S.F.; Riza, N.A. General formula for coupling-loss characterization of single-mode fiber collimators by use of gradient-index rod lenses. *App. Opt.* **1999**, *38*, 3214–3222. [CrossRef]
16. Liu, H.Z.; Liu, L.R.; Xu, R.W.; Luan, Z. ABCD matrix for reflection and refraction of Gaussian beams at the surface of a parabola of revolution. *App. Opt.* **2005**, *44*, 4809–4813. [CrossRef]

MDPI

Article

Research on a Dual-Mode Infrared Liquid-Crystal Device for Simultaneous Electrically Adjusted Filtering and Zooming

Zhonglun Liu [1,2], Mingce Chen [2,3], Zhaowei Xin [2,3], Wanwan Dai [2,3], Xinjie Han [2,3], Xinyu Zhang [2,3,4,*], Haiwei Wang [4] and Changsheng Xie [4]

1 China-EU Institute for Clean and Renewable Energy, Huazhong University of Science & Technology, Wuhan 430074, China; liuzhonglun@hust.edu.cn
2 National Key Laboratory of Science and Technology on Multispectral Information Processing, Huazhong University of Science & Technology, Wuhan 430074, China; mclovelz@hust.edu.cn (M.C.); D201577548@hust.edu.cn (Z.X.); M201672381@hust.edu.cn (W.D.); M201672398@hust.edu.cn (X.H.)
3 School of Automation, Huazhong University of Science & Technology, Wuhan 430074, China
4 Wuhan National Laboratory for Optoelectronics, Huazhong University of Science & Technology, Wuhan 430074, China; hiway@hust.edu.cn (H.W.); Cs_xie@hust.edu.cn (C.X.)
* Correspondence: x_yzhang@hust.edu.cn

Received: 12 December 2018; Accepted: 13 February 2019; Published: 19 February 2019

Abstract: A new dual-mode liquid-crystal (LC) micro-device constructed by incorporating a Fabry–Perot (FP) cavity and an arrayed LC micro-lens for performing simultaneous electrically adjusted filtering and zooming in infrared wavelength range is presented in this paper. The main micro-structure is a micro-cavity consisting of two parallel zinc selenide (ZnSe) substrates that are pre-coated with ~20-nm aluminum (Al) layers which served as their high-reflection films and electrodes. In particular, the top electrode of the device is patterned by 44 × 38 circular micro-holes of 120 μm diameter, which also means a 44 × 38 micro-lens array. The micro-cavity with a typical depth of ~12 μm is fully filled by LC materials. The experimental results show that the spectral component with needed frequency or wavelength can be selected effectively from incident micro-beams, and both the transmission spectrum and the point spread function can be adjusted simultaneously by simply varying the root-mean-square value of the signal voltage applied, so as to demonstrate a closely correlated feature of filtering and zooming. In addition, the maximum transmittance is already up to ~20% according the peak-to-valley value of the spectral transmittance curves, which exhibits nearly twice the increment compared with that of the ordinary LC-FP filtering without micro-lenses.

Keywords: dual-mode liquid-crystal (LC) device; infrared Fabry–Perot (FP) filtering; LC micro-lenses controlled electrically

1. Introduction

In recent years, the infrared detection technology [1] has been developed rapidly, because it presents several advantages including relatively long working distance, better anti-interference, full-time availability, and strong penetration of haze smoke or dust. So far, it has played a significant role in many applications such as the earth resources survey [2], global pollution and disaster warning [3–5], biomedical diagnosis [6,7], food safety monitoring [8–12], and so on. As demonstrated, the electromagnetic radiation out from targets exhibits its own specific spectral characters, which contain objective surface or structural information or chemical composition. Unfortunately, the spectral clues acquired by us are not only originated from targets, but also from environmental circumstances, which means that a mixed broad spectrum can be expected. So, it is of great importance to select only the needed spectral components from incoming beams.

Recently, many methods including typical Mach–Zehnder interferometer [13], Michelson interferometer [14], Bragg grating [15], electro-optical or acousto-optical tuning [16], and Fabry–Perot (FP) micro-cavity [17–20] were constructed successfully. Among them, the FP micro-cavity based on the common interference theory had been developed quickly due to its typical characteristics of ultra-small size and relatively low cost and very high spectral resolution. Combining with the micro-electro-mechanical system (MEMS) [21,22], the electrically tunable MEMS-FP filter had already been effectively constructed. According to the MEMS architecture developed, the depth of the FP micro-cavity could be altered through moving mechanical parts, which supported the mirrors to achieve the spectral choice or further spectral adjustment. At the same time, the liquid-crystal (LC)-FP filters [23–26] composed of the similar FP micro-cavity filled by LC materials had also attracted much attention. As shown, the depth of the FP micro-cavity was fixed, which indicated that the choice or adjustment of the spectral component needed could be conducted by only varying the equivalent index of refraction of the LC layer without any mechanical movement. As known, the characteristics of the lenses with variable focus, as core parts in modern imaging equipment, would greatly determine the final performances of the imaging system. So, the electrically controlled LC micro-lenses [27–32] can be a perfect type of micro-device for flexibly readjusting the patterned light-fields compressed over the surface of the sensor array by main optical system, and thus applied to advanced light-field cameras [33], autostereoscopic devices [34–37] and wave-front measurement and correction sensors [38–40].

In our previous work, a graphene-based LC micro-lens array with a tunable focal length had been developed successfully [41,42]. In order to develop a new kind of dual-mode LC micro-device for simultaneous electrically adjusted filtering and zooming, a structured FP cavity was incorporated into a common LC micro-lens array. The experimental results showed that the maximum transmittance of the spectral micro-beams was already up to ~20% according the peak-to-valley value of the spectral transmittance curves, which exhibited nearly twice the increment compared with the ordinary LC-FP filter. A needed zooming performance was acquired, as shown in the near infrared (NIR) band. When the root-mean-square (RMS) value of the signal voltage was varied from ~2 V_{rms} to ~16 V_{rms}, the point spread function (PSF) of the micro-lenses can be adjusted efficiently, and thus the focal length altered from ~2.80 mm to ~3.93 mm.

2. Materials and Methods

2.1. Device Design and Fabrication

The basic micro-structure of the dual-mode LC micro-device is shown in Figure 1. Figure 1a exhibits the main architecture. Both the patterned Al electrode and the appearance of the actual micro-device are shown in Figure 1b,c, respectively. In the micro-device, ZnSe were chosen as the substrates according to the design of the transmittance being more than 70% in infrared region and having a good tolerance at a relatively high temperature for fabrication process. Firstly, a ~20-nm aluminum (Al) film as a high-reflection film and electrode was simultaneously coated over the surface of ZnSe substrate, since it presented a high reflectivity in the infrared region and a nice electrical conductivity. The top electrode of the micro-device was patterned by 44 × 38 circular micro-holes of 120 μm diameter, which also meant a 44 × 38 micro-lens array. The center-to-center distance between adjacent micro-holes was 336 μm. An alignment layer for initially directing LC molecules made of a thin film of polyamide (PI) was firstly coated on the top and bottom Al film, respectively, and then rubbed in anti-parallel direction to form V-grooves. Both ZnSe substrates were placed in parallel to each other and separated by glass micro-spheres with a typical diameter of ~12 μm so as to construct a stable micro-cavity, where a layer of nematic LC materials of Merck E44 (n_o = 1.5280, n_e = 1.7904, $K_{11} = 15.5 \times 10^{-6}$, $K_{22} = 13.0 \times 10^{-6}$, $K_{33} = 28.0 \times 10^{-6}$, $ep = 22\varepsilon_0$, $et = 5.2\varepsilon_0$) was fully filled. The outline size of the final micro-cavity fabricated was about 3 cm × 2 cm × 2 cm.

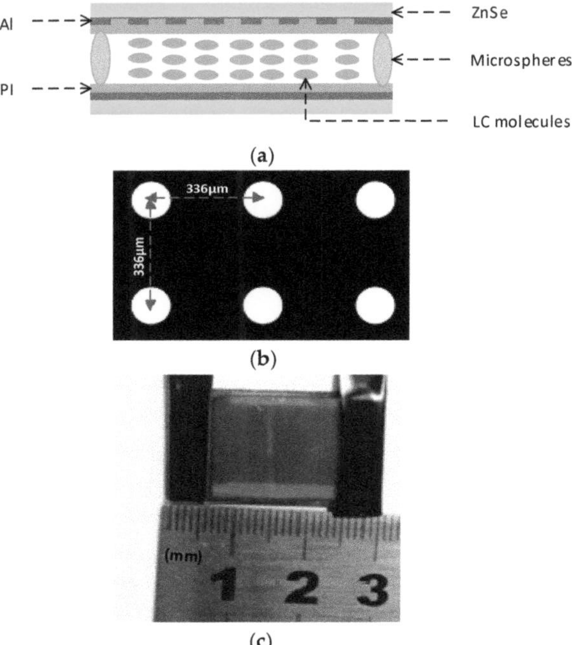

Figure 1. The dual-mode liquid-crystal (LC) micro-device—(**a**) main architecture; (**b**) patterned Al electrode; and (**c**) appearance of the final micro-device.

2.2. Theoretical Analysis

According to the design, the multi-interference based on FP effect was performed between the micro-holes or LC micro-lenses. Equation (1) shows a wavelength selection effect, and a relationship between the wavelength and the corresponding transmittance is also revealed in Equation (2), when the incident beams are perpendicular to the micro-device.

$$m\lambda = 2ndcos\alpha \tag{1}$$

where m is a positive integer, λ is the wavelength of the beams satisfying the resonance condition, n is the equivalent refractive index of LC materials, α is the incident angle of incoming beams, and d is the depth of the micro-cavity.

$$T(\lambda) = (1 - \frac{A}{1-R})^2 \frac{1}{1 + \frac{4R}{(1-R)^2}\sin^2(\frac{2\pi nd}{\lambda} + \varphi)} \tag{2}$$

where T is the transmittance of the micro-device, A is the absorptivity of all materials used in fabricating the micro-device, R is the reflectivity of Al film, and φ is the phase shift of the beams.

If the incident beams are perpendicularly passing through the micro-device, $\alpha = 0$, then only needed spectral micro-beams in a specific wavelength narrow band and satisfying the condition according to Equation (1) can emit at a high transmittance, so as to realize desired filtering operation. Generally, there are two ways to adjust the emitting wavelength—one is to vary the depth of the cavity, as shown in the MEMS-FP; the other is to alter only the refractive index of the medium between the electrodes of the micro-device, which is exactly the situation proposed by us. Since the LC ordering was influenced not only by surface anchoring but also by bulk elastic energy [43,44], a key parameter d

was introduced to represent the micro-device and an experienced value of 12 μm was used to shape the functional LC layer for exhibiting both the zooming function of the LC micro-lens and the filtering function of the FP cavity, simultaneously.

When an electric field was stimulated effectively between the top and bottom electrodes, LC molecules distributed over the surface of the micro-hole would stay in their initial constraint state according to the appearance of the shaped V-grooves, because no electric field can be shaped at the center of each micro-hole. In other words, outside the micro-holes or in the effective FP region, the electric field was perpendicular to the electrode, and most of LC molecules were driven to align along the electric field lines except for partial LC molecules directly contacting with PI layer of the top electrode, and thus strongly directed by PI fabricated V-grooves. It should be noted that the closer to the central area of the micro-hole, the greater the distortion of the electric field lines, and the larger the inclination angle of LC molecules distributed over the middle LC layer corresponding to each LC micro-lens. In addition, the inclination can also be related to the intensity of the electric field stimulated. Generally, it would increase when the electric field was enhanced. The equivalent indices of LC materials varied with the change of the inclination angle of LC molecules, as revealed in Equation (3).

$$n = \frac{n_o n_e}{\sqrt{{n_e}^2 \cos^2\theta + {n_o}^2 \sin^2\theta}} \tag{3}$$

where θ is the inclination angle, n_o and n_e are the refractive indices of LC materials corresponding to the ordinary light and the extraordinary light, respectively. It can be seen that when the stimulated electric field was enhanced, the inclination of LC molecules became larger and then the equivalent index of refraction was decreased. As a result of the reduction of the stimulated electric field, the inclination would be small and thus resulted in an increase of the equivalent index of refraction. Therefore, the refractive indices of LC materials would decrease gradually from the center with respect to the micro-hole to the edge, so as to lead to a needed decrease of light path. Eventually, the beam converging effect appears. Furthermore, since both the transmission spectrum and the focal length of the micro-device are obviously related with the equivalent index of the LC layer or the inclination of the LC molecules, they will be varied simultaneously with the alteration of the signal voltage applied.

3. Experiments and Results

During measurements, a Fourier transform infrared spectrometer (FTIR) of EQUINOX 55 (Bruker, Madison, WI, USA) was used to test the transmission spectrum. The sampling rate was about 80 spectra per second, and the wave numbers were from 4000 cm^{-1} to 400 cm^{-1}, which meant that the wavelength range was from 2.5 μm to 25 μm. A square wave signal with a 1000 Hz frequency and a duty ratio of 1:1 was applied, and its voltage increased gradually from 0 V_{rms} to 22 V_{rms}. Figure 2 reveals the measurement results about spectral filtering. The transmission of ZnSe substrate covered with Al film is shown in Figure 2a. Figure 2b–d show the transmittance spectra of the dual-mode LC micro-device in several wavelength bands including ~2.5 to ~3.2 μm, ~3.6 to ~5.2 μm, and ~10.5 to ~12.0 μm, at the signal voltage of 0 V_{rms}, 4.01 V_{rms}, 7.99 V_{rms}, 10.02 V_{rms}, 16.02 V_{rms}, and 22.00 V_{rms}.

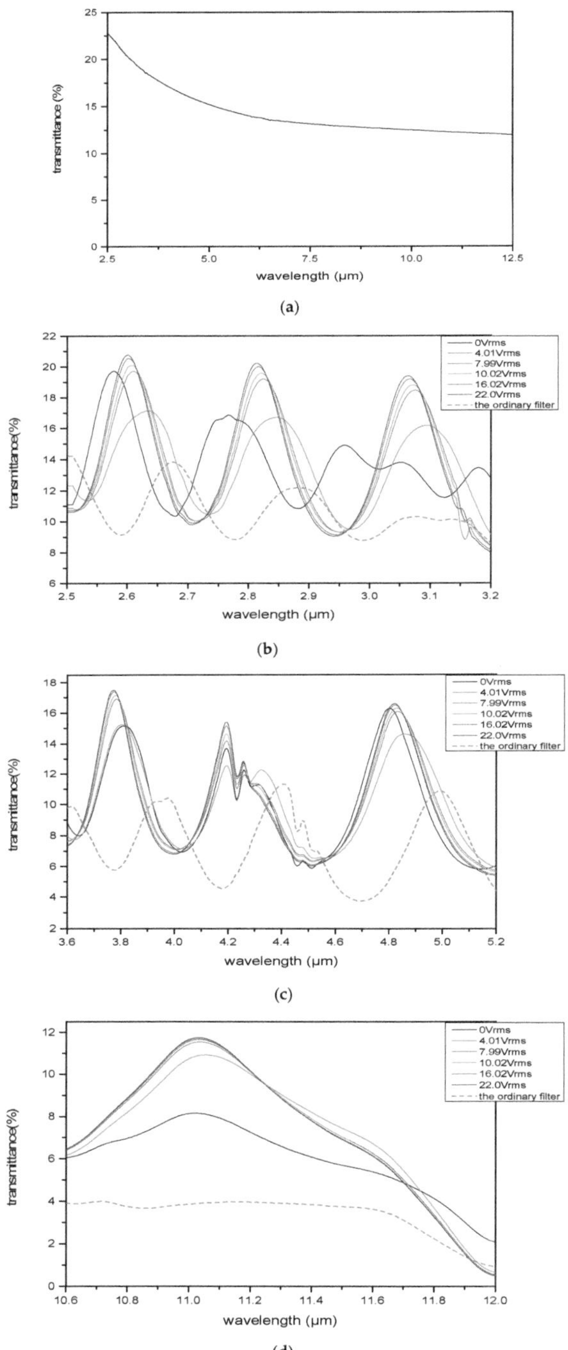

Figure 2. The results of spectral filtering showed the transmittance of (**a**) the substrate covered by Al film; (**b**) wavelength band from 2.5 to 3.2 μm; (**c**) wavelength band from 3.6 to 5.2 μm; and (**d**) wavelength band from 10.6 to 12.0 μm. The signal voltages included were 0 V_{rms}, 4.01 V_{rms}, 7.99 V_{rms}, 10.02 V_{rms}, 16.02 V_{rms}, and 22.00 V_{rms}.

As shown, the transmission of the substrate coated with Al film decreased from ~22.5% to ~12.5% with the increase of the wavelength from ~2.5 to ~12.5 μm, and the average value was about 15%. So, the reflectance of the micro-mirror was about 85%. The spectra indicated by solid lines were the performances of the dual-mode LC micro-device. Since LC of E44 presented a strong absorption in the wavelength range ~3.2 to ~3.6 μm, the wavelength bands of ~2.5 to ~3.2 μm and ~3.6 to ~5.2 μm were analyzed carefully. It can be clearly seen that there were three peaks in each band, and the maximum peak transmittance was up to ~20%. The light loss was mainly due to the strong absorption of Al film and ZnSe and LC materials. As shown, an effective transmission peak with a maximum value of ~12% appeared in the long-wave infrared. When the RMS value of the signal voltage was varied, the transmission spectrum was also shifted, so as to prove the electrically adjusting feature of the developed micro-device. As demonstrated, the maximum adjusting range had already extended to ~80 nm. Furthermore, the dotted lines in the pictures represent the spectrum of the ordinary LC-FP filter without micro-lenses at 0 V_{rms}. Obviously, the transmittance of the micro-device proposed was around 5% higher at the same voltage. In addition, the peak-to-valley value of the obtained spectrum exhibited nearly twice the increment compared with the ordinary one, which was a new result.

The measurement platform for acquiring zooming performances is exhibited in Figure 3. Figure 3a shows the schematic diagram of the optical path, and Figure 3b displays the actual system. During measurements, an infrared beam was emitted from a NIR laser with a central wavelength of ~980 nm, and then weakened by two polarizers, and finally a vertical polarization component being into the micro-device and continuously received by a beam profiler with a magnifying objective of 10×.

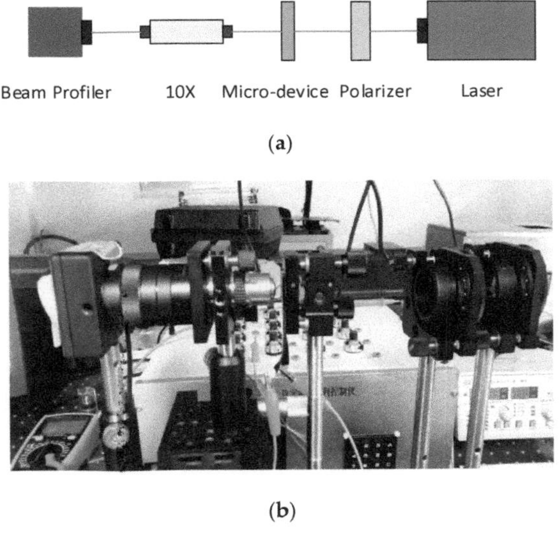

Figure 3. Optical measurement platform for zooming performance—(**a**) schematic diagram and (**b**) actual testing platform.

In experiments, the applied signal voltage was kept constant, for example, at ~4 V_{rms}, and then the distance between the objective lens and the center of the micro-device was adjusted continuously. Figure 4 shows the light intensity distribution formed by LC micro-lenses. The 2D and 3D light intensity distributions at 0 V_{rms} corresponding to a distance of ~3.550 mm are displayed in Figure 4a,b. Figure 4c–e exhibit the 2D light intensity distribution at ~2.215 mm, ~2.555 mm, and ~3.550 mm at ~4 V_{rms}. Figure 4d displays a 3D light intensity distribution at the distance of ~3.550 mm at ~4 V_{rms}. Testing results suggested that the micro-device demonstrated a needed micro-beam convergence performance when an electric field was applied, and the facula size of each LC micro-lens decreased gradually with the distance increasing from ~2.215 mm to ~3.550 mm at ~4 V_{rms}. Since the PSF was already centralized and sharpened fully at ~3.550 mm, this value can be taken as the focal length of the micro-device at ~4 V_{rms}. By changing the applied signal voltage and repeating the measurements, the focal length of the LC micro-device under different voltages can be acquired. The relationship between the focal length and the applied signal voltage is indicated in Figure 5. When the voltage was 2 V_{rms}, the focal length was ~3.93 mm and then decreased with the increase of the voltage. As shown, the minimum value of ~2.80 mm was at 8 V_{rms} and then increased gradually to ~3.50 mm at ~16 V_{rms}. Beyond this working range, the focusing performance was poor because the electric field in the LC micro-lenses was not suitable for forming an effective gradient distribution and variance of the refractive index.

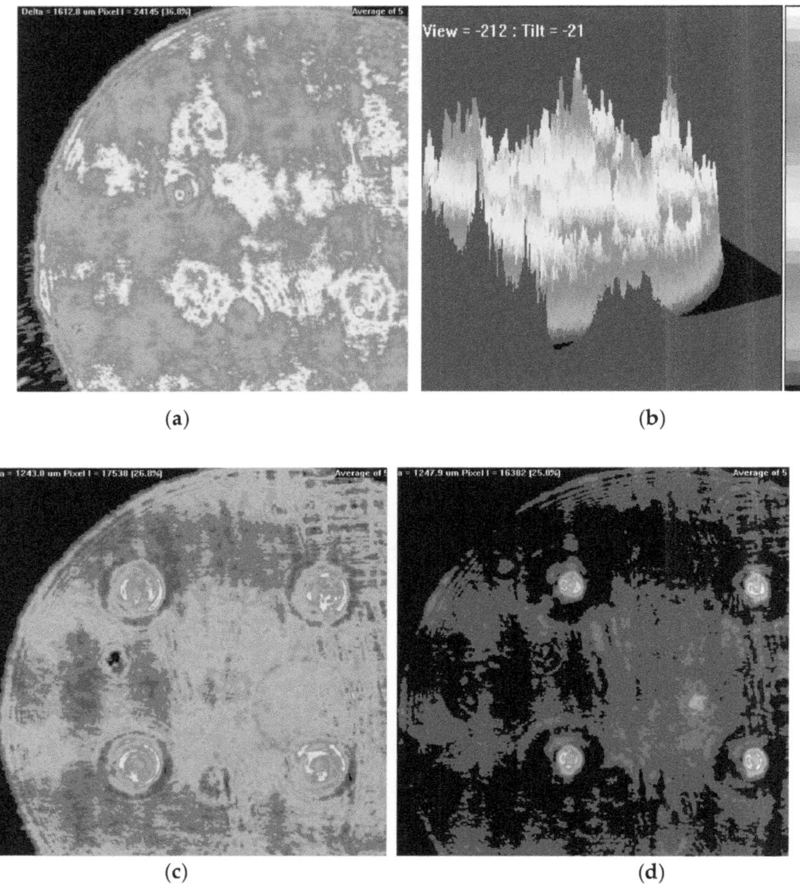

Figure 4. *Cont.*

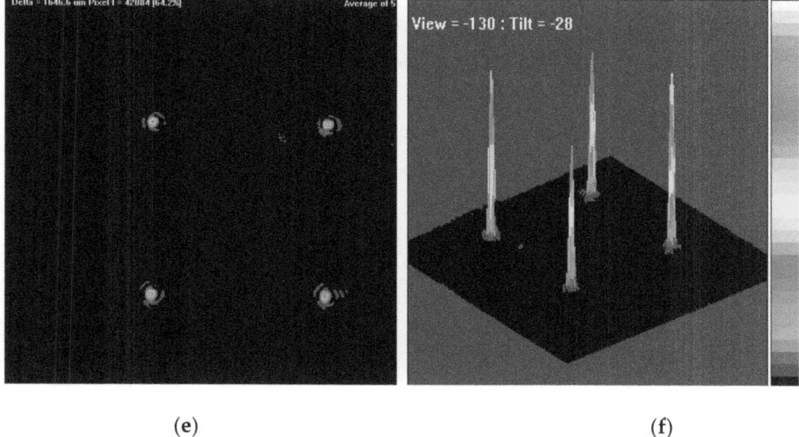

Figure 4. The light intensity distribution formed by LC micro-lenses. (**a**) The 2D light intensity distribution at 0 V_{rms} at the distance of ~3.550 mm and (**b**) the 3D light intensity distribution at 0 V_{rms} at the distance of ~3.550 mm. The 2D light intensity distribution at ~4 V_{rms} at the distance of (**c**) ~2.215 mm, (**d**) ~2.555 mm, and (**e**) ~3.550 mm. (**f**) The 3D light intensity distribution at ~4 V_{rms} at the distance of ~3.550 mm.

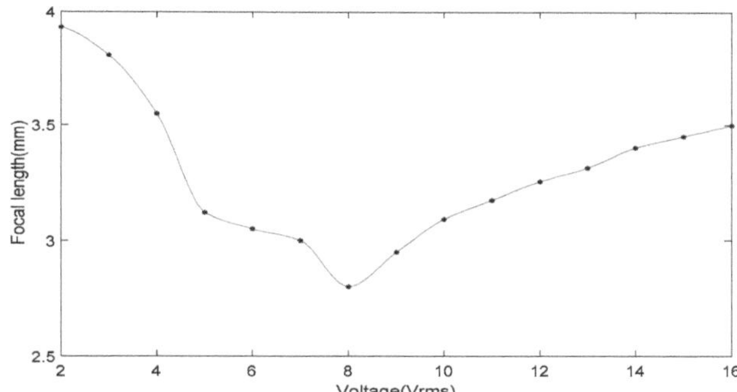

Figure 5. The relationship between the focal length of the dual-mode LC micro-device and the signal voltage applied.

Figure 6 shows a cross-sectional view of the structural piece of the dual-mode LC micro-device. The working area of each LC micro-lens includes region -I, -II, and -III, and the width of the region -I or -III, as shown, is ~20% of that of region -II or the diameter of a micro-hole. Moreover, when the light is incident upon the surfaces of the region -I and -III, the emitted beams should be in a narrow wavelength band, due to the micro-beam filtering effect of the FP cavity constructed, and finally resulted in a slight weakness of the micro-beam intensity converged by the LC micro-lens. For the FP cavity, the main working area is region -IV, and considering the superposition of light going through region -I, -II, and -III, the final spectral transmittance will be increased, as shown in Figure 2.

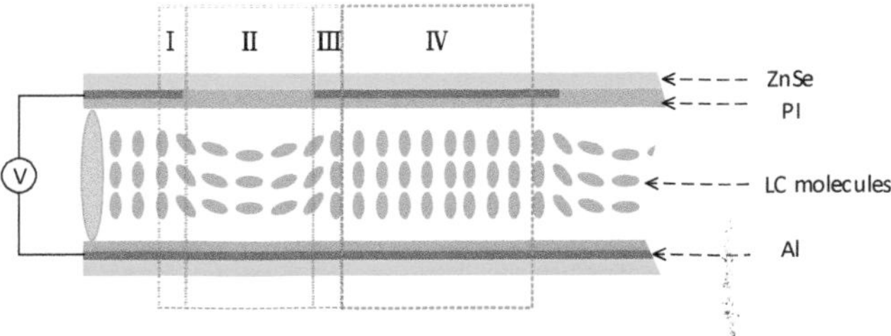

Figure 6. The cross-sectional view of the structural piece of the dual-mode LC micro-device.

From the experiments, it was clear that the proposed micro-device had some special features derived from a FP filter and an arrayed LC micro-lens, because the filtering and zooming operation were closely correlated. When varying the RMS value of the signal voltage applied, the transmission spectrum and focal length of the micro-device were adjusted simultaneously, so as to demonstrate a featured spectral function of the dual-mode micro-device different from common FP filters and LC micro-lenses coupled with sensors. In addition, the obtained infrared spectrum involved several small wavelength bands. The focusing beams shaped at the sensor array also showed a relatively narrow wavelength region corresponding to incident infrared radiation. Through changing the applied electric filed over the functional LC layer, the filtered wave bands were varied. Then, the wave bands of focusing beams at the surface of the sensor array were also affected by changing the focused spectral components, which also involved micro-beams from the region I and III, so as to demonstrate a complicated behavior of adjusting spectrum. Currently, the relationship between filtered wave bands and focused wave bands is being studied carefully.

4. Conclusions

In this paper, a new kind of dual-mode LC micro-device for simultaneously electrically controlled filtering and zooming in infrared band was proposed. The testing results showed that the spectral component with needed light frequency or wavelength band can be effectively selected from incident micro-beams, and the transmission spectrum and the zooming performance were able to be adjusted simultaneously by varying only the RMS value of the signal voltage applied over the micro-device. The correlation relationship between the filtered wave bands and the focused wave bands should be discussed carefully. Due to the ability of recording the light field information of targets by LC micro-lenses, the dual-mode micro-device has a good application prospect in integrating it into infrared light field cameras for conveniently conducting digital refocusing and three-dimensional pattern reconstructing.

Author Contributions: X.Z. conceived the idea; Z.L., M.C., and Z.X. designed and performed the experiments; W.D. and X.H. discussed the results; X.Z., H.W., and C.X. supervised the project; Z.L. wrote the original draft; and X.Z. reviewed the final manuscript.

Funding: This work was supported by the National Nature Science Foundation of China (Nos. 61432007 and 61176052), the Major Technological Innovation Projects in Hubei Province (No. 2016AAA010), and the China Aerospace Science and Technology Innovation Fund (CASC2015).

Acknowledgments: The authors would like to thank the Analytical and Testing Center of Huazhong University of Science and Technology for their valuable help.

Conflicts of Interest: The authors declare no conflict of interest.

References

1. Choi, K.K.; Allen, S.C.; Sun, J.G.; Wei, Y.; Olver, K.A.; Fu, R.X. Resonant structures for infrared detection. *Appl. Opt.* **2017**, *56*, B26–B36. [CrossRef] [PubMed]
2. Kana, J.D.; Djongyang, N.; Raïdandi, D.; Nouck, P.N.; Dadjé, A. A review of geophysical methods for geothermal exploration. *Renew. Sustain. Energy Rev.* **2015**, *44*, 87–95. [CrossRef]

3. Safieddine, S.; Boynard, A.; Coheur, P.-F.; Hurtmans, D.; Pfister, G.; Quennehen, B.; Thomas, J.-C.; Raut, J.-C.; Law, K.S.; Klimont, Z. Summertime tropospheric ozone assessment over the Mediterranean region using the thermal infrared IASI/MetOp sounder and the WRF-Chem model. *Atmos. Chem. Phys.* **2014**, *14*, 10119–10131. [CrossRef]
4. Tagg, A.S.; Sapp, M.; Harrison, J.P.; Ojeda, J.J. Identification and quantification of microplastics in wastewater using focal plane array-based reflectance micro-FT-IR imaging. *Anal. Chem.* **2015**, *87*, 6032–6040. [CrossRef]
5. Li, F.; Yang, W.; Liu, X.; Sun, G.; Liu, J. Using high-resolution UAV-borne thermal infrared imagery to detect coal fires in Majiliang mine, Datong coalfield, Northern China. *Remote Sens. Lett.* **2018**, *9*, 71–80. [CrossRef]
6. Kong, K.; Kendall, C.; Stone, N.; Notingher, I. Raman spectroscopy for medical diagnostics—from in-vitro biofluid assays to in-vivo cancer detection. *Adv. Drug Delivery Rev.* **2015**, *89*, 121–134. [CrossRef] [PubMed]
7. Pilling, M.; Gardner, P. Fundamental developments in infrared spectroscopic imaging for biomedical applications. *Chem. Soc. Rev.* **2016**, *45*, 1935–1957. [CrossRef]
8. Scholl, P.F.; Bergana, M.M.; Yakes, B.J.; Xie, Z.; Zbylut, S.; Downey, G.; Buehler, M. Effects of the adulteration technique on the near-infrared detection of melamine in milk powder. *J. Agric. Food Chem.* **2017**, *65*, 5799–5809. [CrossRef]
9. Jaiswal, P.; Jha, S.N.; Kaur, J.; Borah, A. Detection and quantification of anionic detergent (lissapol) in milk using attenuated total reflectance-Fourier Transform Infrared spectroscopy. *Food Chem.* **2017**, *221*, 815–821. [CrossRef]
10. Qu, J.H.; Liu, D.; Cheng, J.H.; Sun, D.W.; Ma, J.; Pu, H.; Zeng, X.A. Applications of near-infrared spectroscopy in food safety evaluation and control: A review of recent research advances. *Crit. Rev. Food Sci. Nutr.* **2015**, *55*, 1939–1954. [CrossRef]
11. Kamruzzaman, M.; Makino, Y.; Oshita, S. Rapid and non-destructive detection of chicken adulteration in minced beef using visible near-infrared hyperspectral imaging and machine learning. *J. Food Eng.* **2016**, *170*, 8–15. [CrossRef]
12. Huang, M.; Kim, M.S.; Delwiche, S.R.; Chao, K.; Qin, J.; Mo, C.; Esquerre, C.; Zhu, Q. Quantitative analysis of melamine in milk powders using near-infrared hyperspectral imaging and band ratio. *J. Food Eng.* **2016**, *181*, 10–19. [CrossRef]
13. Kumar, S.; Singh, G.; Bisht, A. 4 $\times$ 4 signal router based on electro-optic effect of Mach–Zehnder interferometer for wavelength division multiplexing applications. *Opt. Commun.* **2015**, *353*, 17–26. [CrossRef]
14. Harlander, J.M.; Englert, C.R.; Brown, C.M.; Marr, K.D.; Miller, L.J.; Zastera, V.; Bach, B.W.; Mende, S.B. Michelson interferometer for global high-resolution thermospheric imaging (MIGHTI): Monolithic interferometer design and test. *Space Sci. Rev.* **2017**, *212*, 601–613. [CrossRef] [PubMed]
15. Campanella, C.E.; De Leonardis, F.; Mastronardi, L.; Mastronardi, L.; Malara, P.; Gagliardi, G.; Passaro, V.M. Investigation of refractive index sensing based on Fano resonance in fiber Bragg grating ring resonators. *Opt. Express* **2015**, *23*, 14301–14313. [CrossRef] [PubMed]
16. Plascak, M.E.; Ramirez, R.B.; Bagnell, K.; Delfyett, P.J. Tunable Broadband Electro-Optic Comb Generation Using an Optically Filtered Optoelectronic Oscillator. *IEEE Photonics Technol. Lett.* **2018**, *30*, 335–338. [CrossRef]
17. Plumb, A.A.; Huynh, N.T.; Guggenheim, J.; Zhang, E.; Beard, P. Rapid volumetric photoacoustic tomographic imaging with a Fabry-Perot ultrasound sensor depicts peripheral arteries and microvascular vasomotor responses to thermal stimuli. *Eur. Radiol.* **2018**, *28*, 1037–1045. [CrossRef] [PubMed]
18. Barnes, J.; Li, S.; Goyal, A.; Abolmaesumi, P.; Mousavi, P.; Loock, H.P. Broadband Vibration Detection in Tissue Phantoms Using a Fiber Fabry–Perot Cavity. *IEEE Trans. Biomed. Eng.* **2018**, *65*, 921–927. [CrossRef]
19. Islam, M.; Ali, M.; Lai, M.H.; Lim, K.S.; Ahmad, H. Chronology of Fabry-Perot interferometer fiber-optic sensors and their applications: A review. *Sensors* **2014**, *14*, 7451–7488. [CrossRef]
20. Wang, N.; Li, J.; Wei, G.; Talbi, L.; Zeng, Q.; Xu, J. Wideband Fabry–Perot resonator antenna with two layers of dielectric superstrates. *IEEE Antennas Wirel. Propag. Lett.* **2015**, *14*, 229–232. [CrossRef]
21. Erfan, M.; Sabry, Y.M.; Sakr, M.; Mortada, B.; Medhat, M.; Khalil, D. On-chip micro–electro–mechanical system fourier transform infrared (MEMS FT-IR) spectrometer-based gas sensing. *Appl. Spectrosc.* **2016**, *70*, 897–904. [CrossRef] [PubMed]

22. Meng, Q.; Chen, S.; Lai, J.; Huang, Y.; Sun, Z. Multi-physics simulation and fabrication of a compact 128 × 128 micro-electro-mechanical system Fabry–Perot cavity tunable filter array for infrared hyperspectral imager. *Appl. Opt.* **2015**, *54*, 6850–6856. [CrossRef] [PubMed]
23. Lin, J.; Tong, Q.; Lei, Y.; Xin, Z.; Wei, D.; Zhang, X.; Wang, H.; Xie, C. Electrically tunable infrared filter based on a cascaded liquid-crystal Fabry–Perot for spectral imaging detection. *Appl. Opt.* **2017**, *56*, 1925–1929. [CrossRef] [PubMed]
24. Lin, J.; Tong, Q.; Lei, Y.; Xin, Z.; Zhang, X.; Ji, A.; Sang, H.; Xie, C. An arrayed liquid crystal Fabry–Perot infrared filter for electrically tunable spectral imaging detection. *IEEE Sensors J.* **2016**, *16*, 2397–2403. [CrossRef]
25. Zhang, H.; Muhammad, A.; Luo, J.; Tong, Q.; Lei, Y.; Zhang, X.; Sang, H.; Xie, C. Electrically tunable infrared filter based on the liquid crystal Fabry–Perot structure for spectral imaging detection. *Appl. Opt.* **2014**, *53*, 5632–5639. [CrossRef] [PubMed]
26. Zhang, H.; Muhammad, A.; Luo, J.; Tong, Q.; Lei, Y.; Zhang, X.; Sang, H.; Xie, C. MWIR/LWIR filter based on Liquid–Crystal Fabry–Perot structure for tunable spectral imaging detection. *Appl. Opt.* **2015**, *69*, 68–73. [CrossRef]
27. Urruchi, V.; Algorri, J.F.; Sánchez-Pena, J.M.; Bennis, N.; Geday, M.A.; Otón, J.M. Electrooptic characterization of tunable cylindrical liquid crystal lenses. *Mol. Cryst. Liq. Cryst.* **2012**, *553*, 211–219. [CrossRef]
28. Algorri, J.F.; Urruchi, V.; Bennis, N.; Morawiak, P.; Sánchez-Pena, J.M.; Otón, J.M. Liquid crystal spherical microlens array with high fill factor and optical power. *Opt. Express* **2017**, *25*, 605–614. [CrossRef]
29. Algorri, J.F.; Bennis, N.; Herman, J.; Kula, P.; Urruchi, V.; Sánchez-Pena, J.M. Low aberration and fast switching microlenses based on a novel liquid crystal mixture. *Opt. Express* **2017**, *25*, 14795–14808. [CrossRef]
30. Algorri, J.F.; Urruchi, V.; Bennis, N.; Sánchez Pena, J.M. Using an analytical model to design liquid crystal microlenses. *IEEE Photonics Technol. Lett.* **2014**, *26*, 793–796. [CrossRef]
31. Naumov, A.F.; Love, G.D.; Loktev, M.Y.; Vladimirov, F.L. Control optimization of spherical modal liquid crystal lenses. *Opt. Express* **1999**, *4*, 344–352. [CrossRef] [PubMed]
32. Kirby, A.K.; Hands, P.J.; Love, G.D. Liquid crystal multi-mode lenses and axicons based on electronic phase shift control. *Opt. Express* **2007**, *15*, 13496–13501. [CrossRef] [PubMed]
33. Algorri, J.F.; Urruchi, V.; Bennis, N.; Morawiak, P.; Sánchez-Pena, J.M.; Otón, J.M. Integral imaging capture system with tunable field of view based on liquid crystal microlenses. *IEEE Photonics Technol. Lett.* **2016**, *28*, 1854–1857. [CrossRef]
34. Algorri, J.F.; Urruchi, V.; García-Cámara, B.; Sánchez-Pena, J.M. Liquid crystal microlenses for autostereoscopic displays. *Materials* **2016**, *9*, 36. [CrossRef] [PubMed]
35. Algorri, J.F.; Urruchi, V.; Sánchez-Pena, J.M.; Otón, J.M. An autostereoscopic device for mobile applications based on a liquid crystal microlens array and an OLED display. *J. Disp. Technol.* **2014**, *10*, 713–720. [CrossRef]
36. Algorri, J.F.; Urruchi, V.; Bennis, N.; Sánchez-Pena, J.M.; Otón, J.M. Cylindrical liquid crystal microlens array with rotary optical power and tunable focal length. *IEEE Electron Device Lett.* **2015**, *36*, 582–584. [CrossRef]
37. Lee, Y.H.; Peng, F.; Wu, S.T. Fast-response switchable lens for 3D and wearable displays. *Opt. Express* **2016**, *24*, 1668–1675. [CrossRef] [PubMed]
38. Loktev, M.Y.; Belopukhov, V.N.; Vladimirov, F.L.; Vdovin, G.V.; Love, G.D.; Naumov, A.F. Wave front control systems based on modal liquid crystal lenses. *Rev. Sci. Instrum.* **2000**, *71*, 3290–3297. [CrossRef]
39. Hands, P.J.; Tatarkova, S.A.; Kirby, A.K.; Love, G.D. Modal liquid crystal devices in optical tweezing: 3D control and oscillating potential wells. *Opt. Express* **2006**, *14*, 4525–4537. [CrossRef]
40. Algorri, J.F.; Urruchi, V.; Bennis, N.; Sánchez-Pena, J.M.; Otón, J.M. Tunable liquid crystal cylindrical micro-optical array for aberration compensation. *Opt. Express* **2015**, *23*, 13899–13915. [CrossRef]
41. Xin, Z.; Wei, D.; Chen, M.; Hu, C.; Li, J.; Zhang, X.; Liao, J.; Wang, H.; Xie, C. Graphene-based adaptive liquid-crystal microlens array for a wide infrared spectral region. *Opt. Mater. Express* **2019**, *9*, 183–194. [CrossRef]
42. Wu, Y.; Hu, W.; Tong, Q.; Lei, Y.; Xin, Z.; Wei, D.; Zhang, X.; Liao, J.; Wang, H.; Xie, C. Graphene-based liquid-crystal microlens arrays for synthetic-aperture imaging. *J. Opt.* **2017**, *19*, 095102. [CrossRef]
43. Wang, X.; Zhou, Y.; Kim, Y.K.; Miller, D.S.; Zhang, R.; Martinez-Gonzalez, J.A.; Bukusoglu, E.; Zhang, B.; Brown, T.M.; Pablo, J.J. Patterned surface anchoring of nematic droplets at miscible liquid–liquid interfaces. *Soft Matter* **2017**, *13*, 5714–5723. [CrossRef] [PubMed]

44. Miller, D.S.; Wang, X.; Abbott, N.L. Design of functional materials based on liquid crystalline droplets. *Chem. Mater.* **2013**, *26*, 496–506. [CrossRef] [PubMed]

micromachines

Review

Metalens-Based Miniaturized Optical Systems

Bo Li [1,2], Wibool Piyawattanametha [2,3] and Zhen Qiu [1,2,4,*]

1 Department of Electrical and Computer Engineering, Michigan State University, East Lansing, MI 48823, USA; libo2@msu.edu
2 Institute for Quantitative Health Science and Engineering, Michigan State University, East Lansing, MI 48823, USA; wibool@gmail.com
3 Department of Biomedical Engineering, Faculty of Engineering King Mongkut's Institute of Technology Ladkrabang (KMITL), Bangkok 10520, Thailand
4 Department of Biomedical Engineering, Michigan State University, East Lansing, MI 48823, USA
* Correspondence: qiuzhen@egr.msu.edu; Tel.: +1-517-884-6942

Received: 31 March 2019; Accepted: 4 May 2019; Published: 8 May 2019

Abstract: Metasurfaces have been studied and widely applied to optical systems. A metasurface-based flat lens (metalens) holds promise in wave-front engineering for multiple applications. The metalens has become a breakthrough technology for miniaturized optical system development, due to its outstanding characteristics, such as ultrathinness and cost-effectiveness. Compared to conventional macro- or meso-scale optics manufacturing methods, the micro-machining process for metalenses is relatively straightforward and more suitable for mass production. Due to their remarkable abilities and superior optical performance, metalenses in refractive or diffractive mode could potentially replace traditional optics. In this review, we give a brief overview of the most recent studies on metalenses and their applications with a specific focus on miniaturized optical imaging and sensing systems. We discuss approaches for overcoming technical challenges in the bio-optics field, including a large field of view (FOV), chromatic aberration, and high-resolution imaging.

Keywords: metasurface; metalens; field of view (FOV); achromatic; Huygens' metalens; bio-optical imaging; optical coherence tomography; confocal; two-photon; spectrometer

1. Introduction

Miniaturized optical systems, for both imaging and sensing, have recently become very attractive for many biomedical applications, such as wearable and endoscopic medical devices. Novel optical lenses with ultrathin structure and light weight have played an important role in the miniaturization of state-of-the-art bio-optical systems. Traditional planar optical lenses (such as micro-gratings and Fresnel micro-lenses) and thin-film micro-optics have been studied in the last few decades. Although the device's footprint has been slightly reduced by using these lenses, conventional lenses have already been shown to have many disadvantages, including limited optical quality for imaging, integration difficulties, and high cost. Metasurface-based flat optical lenses (so-called metalenses) [1–7] show great potential and could overcome most of the challenges. The meta building blocks (MBBs) work as subwavelength-spaced scatterers. Many basic properties of light [8–10] (such as phase, polarization, and focal points) can be controlled in high-resolution imaging and sensing, through tuning the MBBs' shapes, size, and positions. Conventional lenses, such as refractive lenses (objectives and telescope), are usually bulky and expensive, although they are still dominant in optical systems. Unfortunately, their fabrication processes (such as molding, polishing, and diamond-turning) are commonly sophisticated. In addition, the phase profiles are quite limited, while the structure of the lenses is small. On the contrary, metalenses overcome those limitations and provide great advantages compared to traditional optical elements. Especially by using accurate numerical methods, the phase profiles of metalenses can be well designed with MBBs. With advanced micro-machining processes, metalenses can be mass-produced with high yield.

2. The Fundamentals of Metasurface-Based Lenses

2.1. Phase Profile Control

Refractive lenses are widely used in various optical systems such as telescopes and microscopes. Although they have very good properties in phase control and polarization, traditional refractive lenses with a high numerical aperture (NA) are often bulky and expensive. Additionally, the complex macro- or meso-scale fabrication process still relies on conventional optics manufacturing methods, which have been developed for over 100 years. To meet the optical requirements, refractive lenses are usually designed with different shapes. However, a metalens provides new opportunities to overcome these limitations. For instance, the phase profile can be modified by changing the MBBs [11]. The hyperbolic phase profile [2] required for focusing a normal incident beam that remains collimated inside the substrate can be expressed as follows:

$$\varphi(r) = -\frac{2\pi}{\lambda}\left(\sqrt{r^2 + f^2} - f\right) \tag{1}$$

where f is the focal length of the illumination wavelength and r is the radial coordinate. The designed metasurface should create a phase profile to modulate the incident planar wavefront into spherical ones at focal length f from the lenses.

2.2. Plasmonic Metasurface-Based Lenses

Usually, metasurface-based lenses use MBBs to modify the optical characteristics. One of the most representative techniques is to create plasmonic effects on the surface. A plasmonic antenna [12] can be easily micro-machined using advanced electron beam lithography (EBL) and a relatively simple lift-off process. The concentrated incident light can be transformed into a smaller region that matches its own wavelength and causes oscillations. By having the plasmonic effect on its metasurface, the metalens has attracted great interest in the optics field. For example, experimental results reported by Yin et al. [13] have indicated that the micro-structures have a plasmonic effect on the Ag film surface and successfully formed a focal spot at the focusing plane. Another study, by Zhang et al. [14], has shown that the focusing could be achieved and also tuned using different nano-antenna shapes such as elliptical and circular blocks.

2.3. All Dielectric Metasurface-Based Lenses

While dielectric phase shifters are utilized in the MBBs, the energy absorption loss of the incident could potentially be reduced significantly. Researchers have taken advantage of this and included dielectric phase shifters in many new optics designs. For example, Vo et al. [10] proposed polarization independent lenses with dielectric building blocks (circular silicon arrays). High transmission efficiency (70%) has been achieved with the incident light at a wavelength of 850 nm. Faraon's group from Caltech [15] has demonstrated that a single dielectric nano-antenna could be designed as an efficient building block that might provide full phase coverage. Based on the optimized nanostructures, the spatial image resolution has excellent qualities and relatively high transmission efficiency. Capasso's group from Harvard University [16] has demonstrated that the dielectric metalens also has superior performance in spectral applications in the visible range. The polarization independent metalens has been micro-fabricated by titanium dioxide (TiO_2) nanopillars. The metalens could achieve a relatively high NA = 0.85 with an efficiency of more than 60% for incident wavelengths of 532 nm and 660 nm.

3. Advanced Techniques for Metasurface-Based Lenses

3.1. Wide Angular Field of View

In a miniaturized optical system, the field of view (FOV) is one of the key factors for evaluating the overall qualities of the imaging and sensing system [17]. Unfortunately, due to technical limitations, most metalenses suffer from serious off-axis aberration, leading to a limited FOV. Pursuing increased FOV is a common goal of many scientific studies. For example, in traditional optical lens-based imaging systems, bulky and expensive aberration-corrected objective lenses are frequently utilized to achieve a relatively large FOV.

Theoretically, single-layer metasurface-based flat lenses suffer from off-axis aberrations [18,19], along with wide-angle absorption [20–23] and other problems. To broaden the FOV, multiple-layer metalens structures have been successfully demonstrated. For example, Faraon's group [24] has shown a doublet lens formed by cascading two metasurfaces (Figure 1a), which could achieve limited diffraction, focusing up to ± 30° with near-infrared (NIR) incident light of 850 nm. For a shorter wavelength (532 nm) in the visible range, Capasso's group [25] reported a metalens doublet design. For the aperture metalens, shown in Figure 1b, with positive and negative angle incident light, the spherical aberration can be corrected and all the focusing points can eventually be allocated on the same focal plane. Based on the principle of the Chevalier lens [26], the metalens shown in Figure 1c has provided a relatively larger FOV with an incident light angle up to ±25°.

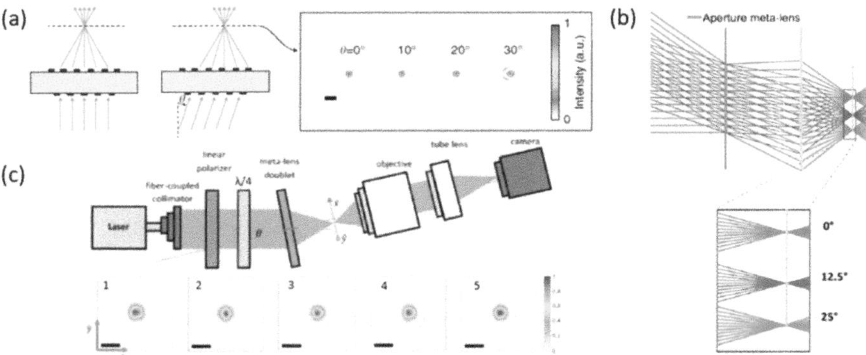

Figure 1. Focal spot characterization for different angles of incidence source. (**a**) (**left**) Schematic illustration of focusing of on-axis and off-axis light by a metasurface doublet lens. (**right**) Simulated focal plane intensity for different incident angles. (Reproduced from [24] with permission.) (**b**) The Ray diagram obtained by adding the aperture meta-lens resulting in diffraction-limited focusing along the focal plane. (**c**) Focal spot measurement setup. (1–5) Focal spot intensity profile at (1) 0°, (2) 6°, (3) 12°, (4) 18°, (5) 25° incidence angle. (Reproduced from [25] with permission.)

Most recently, single-layer metalens with disorder-engineered design [27] providing the optical randomness of conventional disordered media, but in a way that is fully known a priori (Figure 2a), has been demonstrated by Yang and Faraon's groups at Caltech [27] with improved resolution and FOV. This disorder-engineered metalens has individual input-output responses, which is different from the multiple-lens based system. The new metalens has a high numerical aperture (NA ~0.5) focusing to 2.2×10^8 points in an ultra-large FOV with an outer diameter of 8 mm.

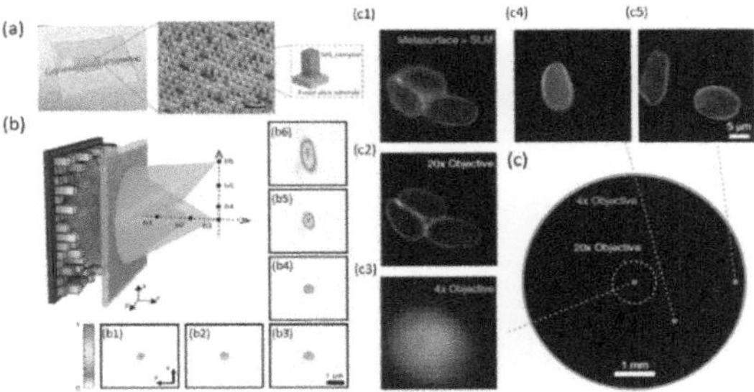

Figure 2. Disorder-engineered metasurfaces. (**a**) Photograph and SEM image of a fabricated disorder-engineered metasurface. (**b**) Schematic of optical focusing assisted by the disordered metasurface. The incident light is polarized along the x direction (b1–b6). (**c**) Low-resolution bright-field image captured by a conventional fluorescence microscope with a 4 × objective lens (NA = 0.1). (Reproduced from [27] with permission.)

To increase the FOV, another approach has been proposed by Guo et al. [28]. The key to constructing a metalens with larger FOV is to realize a perfect conversion from rotational symmetry to translational symmetry in the light field [28]. The wavenumber in free space and incident angles should satisfy the following relation, shown as Equation (2):

$$k_0 sin\theta_x x + k_0 sin\theta_y y + \varnothing_m(x, y) = \varnothing_m\left(x + \Delta_x,\ y + \Delta_y\right) \tag{2}$$

where the k_0 is the wavenumber in free space, $\varnothing_m(x, y)$ is the phase shift profile carried by the flat lens, Δ_x and Δ_y correspond to the translational shift of $\varnothing_m(x, y)$ at incidence angles of θ_x and θ_y [28]. To verify this method, a new metalens with diameter of 350 mm and focal length f = 87.5 mm (NA ~ 0.89) was simulated at 19 GHz, shown in Figure 3a. As a proof of concept, the measurement of far-field power patterns is demonstrated with a circularly polarized horn through the metalens for wide FOV. The result shows that a ± 60° beam steering can be realized by the transversely changing the location of the antenna within an area from − 75.8 mm to + 75.8 mm.

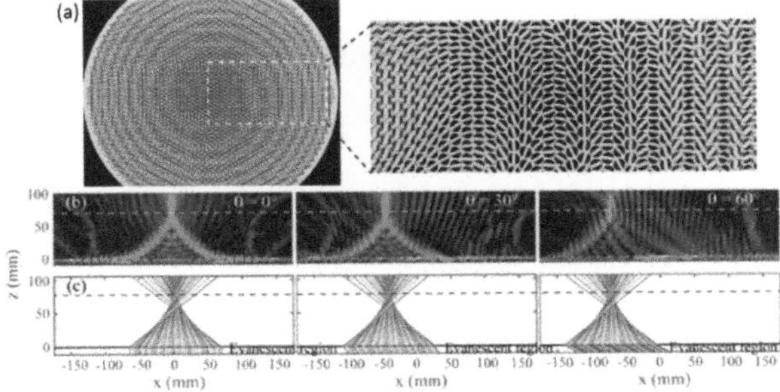

Figure 3. Performance of the wide-angle flat lens. (**a**) Perspective and zoom view of the wide-angle flat lens. (**b**) Simulated light intensity distributed on the *xoz* plane at 19 GHz electric field distributions. (**c**) Ray trajectories of 19 GHz before and after propagating through the flat lens. Left, middle, and right panels of (**b**) and (**c**), respectively, represent the cases of $\varnothing = 0^{\circ}, 30^{\circ}, 60^{\circ}$. (Reproduced from [28] with permission.)

To achieve broader FOV and preserve high-resolution imaging performance, Yang and Faraon's groups [29] have demonstrated a new phase-array-based method (Figure 4) that does not require a large scale-up in the number of controllable elements [29]. It uses disorder-engineered design, as the key factor is similar to that of previous work [27].

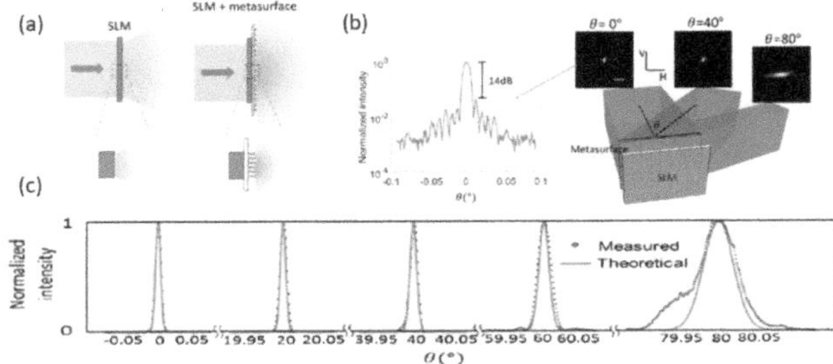

Figure 4. Wide-angular-range and high-resolution beam steering by a metasurface-coupled phased array. (**a**) The comparison of steering range of a single SLM structure and a metasurface-coupled SLM structure. (**left**) without the metasurface, the SLM can provide only a small diffraction envelope. (**right**) With the metasurface-coupled SLM structure, since each scatterer is subwavelength, the steerable range can span from −90° to +90°. (**b**) Illustration of the steering scheme (**c**) 1D far−field beam shapes at other steering angles. Red lines denote the theoretical shapes of the beams. Blue dots denote the measured data. (Reproduced from [29] with permission.)

The 2D array subwavelength scatters (SiN_x with height 630 nm) were deposited on the silica substrate arranged in a square lattice with a pitch size of 350 nm. Also, when the designed lens combined with a spatial light modulator (SLM), the output light of the system would have a larger cover angle range than what is possible with a SLM alone. As a result, the disorder-engineered metasurface with SLM was able to scatter light uniformly within the range of ± 90° (Figure 4a) due to the subwavelength size and random distribution of the nanofins.

Based on the above introduction, the FOV of optical systems can be increased by using proper metasurface design. The nano-element size, material and metasurface structure have different effects to the selected incident light in improving the FOV. A comparison table of different metasurface designs for improving the FOV is shown below (Table 1).

Table 1. Comparison between each fabricated metasurface-based lens in wide angular FOV design.

Reference (Year)	Efficiency	Material	NA	Wavelength	FOV
Arbabi et al. (2016) [24]	70%	a-Si:H	N/A	850 nm	±30°
Groever et al. (2017) [25]	N/A	TiO_2	0.44	532 nm	±25°
Jang et al. (2018) [27]	N/A	SiN_x	>0.5	532 nm	8 mm
Guo et al. (2018) [28]	93%	Simulation	0.89	Far-field power	±60°
Xu et al. (2018) [29]	95%	SiN_x	N/A	532 nm	±80°

3.2. *Achromatic Metasurface-Based Lenses*

With the significant progress in nano-fabrication, researchers not only focus on high-resolution imaging but also exploit the diffractive optics system by overcoming more fundamental problems, such as chromatic aberration. Like the traditional refractive lenses, new techniques have been demonstrated for focusing on multiple different wavelengths' incident light with the same focal length. Traditionally, for apochromatic and super-achromatic lenses [30], macro- or meso-scale manufacturing methods are complex. However, the metasurface provides a new approach because the flexible shape of the electromagnetic field [31] can be modified by changing the phase profile. Capasso's group [32] reported that the phase realization process and interference mechanism result in large chromatic aberrations in diffractive lenses. For the dielectric-based metalens [5], the phase shifter, nanopillars, acts as a truncated waveguide with predetermined dispersion. To realize achromatic metalenses, the key is to optimize the phase shifters' geometric parameters (Figure 5a) by satisfying the phase coverage from 0 to 2π with different wavelength dispersion [33]. The fabricated achromatic metalens (AML), shown in Figure 5b, has more advantages and overcomes the existing optics problems. The primary goal is to maximize the phase coverage. Second, the phase shifters guarantee polarization-insensitive operation for AML. To characterize the performance of the AML, the focal distance for different wavelength incident light was measured (Figure 5c). The simulated focal point intensity is shown in Figure 5d. Even for different wavelength incident light, the focal spots have a perfect shape at the same distance (z = 485 μm). The results show a theoretical and experimental achromatic response, where the focal length remains unchanged with a broad bandwidth (60 nm) in the visible range.

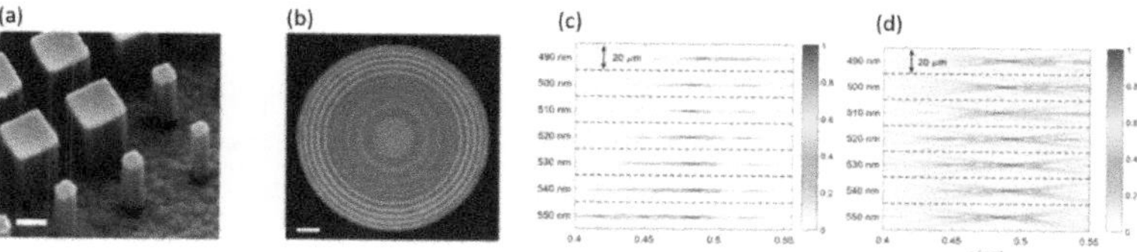

Figure 5. A 60 nm bandwidth achromatic metalens. (**a**) Side view scanning electron microscope (SEM) image of the fabricated AML, scale bar: 200 nm. (**b**) Optical image of the AML. Scale bar: 25 μm. (**c**) Measured intensity profiles of the reflected beam by the AML in the *xz*-plane at different wavelengths. (**d**) Simulated intensity profiles of the reflected beam by the AML in the *xz*-plane at different wavelengths. (Reproduced from [32] with permission.)

When a longer wavelength incident light is used in the miniaturized optical system, it will cause chromatic dispersion [34] because the index of refraction decreases with a longer wavelength. The refractive lenses need to have a larger focal length and prisms, which will deflect at a smaller angle for a longer wavelength. A serious chromatic aberration will degrade the system performance, especially in multi-color imaging applications. Although there have been some good achromatic metalens designs, which can suppress the chromatic effect over 60 nm bandwidth in the visible range [32], the working achromatic bandwidth (~11.4% of the central wavelength) is still not broad enough for practical applications. Wang et al. [35] demonstrated a new broadband achromatic metalens (BAML) (Figure 6) in the infrared (IR) range. The new metalens works within a broad infrared bandwidth at wavelengths from 1200 nm to 1680 nm.

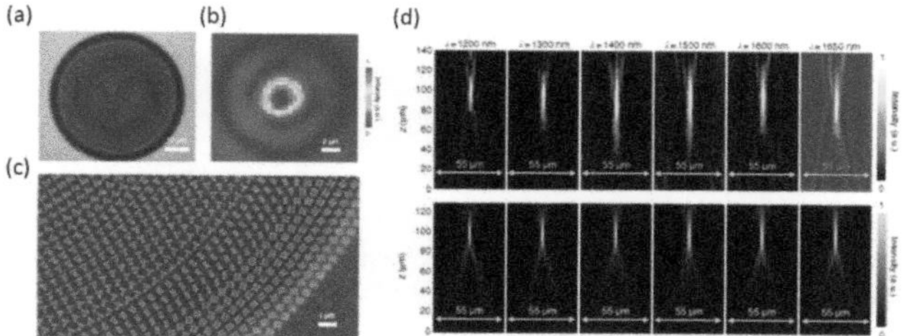

Figure 6. Verification of achromatic converging metalens. (**a**) Optical image of a fabricated metalens with NA = 0.268. (**b**) Measured light intensity of focal spot at incident wavelength λ = 1500 nm. (**c**) Zoomed-in scanning electron microscope (SEM) image of the fabricated metalens. (**d**) Experimental (top row) and numerical (bottom row) intensity profiles of BAML along axial planes at various incident wavelengths. (Reproduced from [35] with permission).

Generally speaking, the dispersion can be divided into two different effects: dispersion elimination and dispersion expansion. For imaging purposes, researchers attempt to eliminate the dispersion because it could cause chromatic aberration and degrade the overall image quality. However, the dispersion can be used to suppress the nonlinear effects during fiber communication [36]. For natural materials, dispersion is determined by their own electronic and energy levels [37]. To resolve this problem, Li et al. [38] from the Chinese Academy of Sciences proposed a new method to control the dispersion. The metalens design was fabricated by silicon nanocuboids, which respond to specific wavelengths (473, 532, and 632.6 nm). With the Pancharatnam-Berry (P-B) phase shift design, the chromatic dispersion among different wavelengths can be engineered independently. Subsequently, a series of flat optical devices with both achromatic and super-dispersive (positive or negative) focusing properties can be demonstrated.

As shown in Figure 7a, there are five designed metasurface lenses. The first metalens M_1 is a flat achromatic lens (f = 10 μm and NA = 0.6294). The schematic and phase profile of M_1 are shown in Figure 7b,c. The second and third metalenses (M_2 and M_3) were designed as super-dispersion metalenses, shown in Figure 7d–g. For M_2, it could focus the red, green, and blue light into different focal lengths (6, 10, and 17 μm). M_3 has a reversed super-dispersion where focal points of red, green, and blue light are in reverse order compared to M_2. M_4 and M_5 are off-axis super-dispersion metalenses, shown in Figure 7h–k).

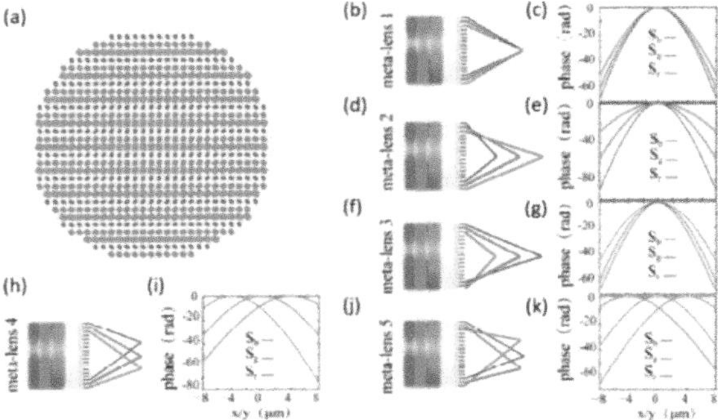

Figure 7. Dispersion controlling meta-lens. (**a**) The schematic of the designed metasurface. (**b**) Schematic and (**c**) phase distribution of achromatic metalens M1. (**d**) Schematic and (**e**) phase distribution of super-dispersion metalens M2 designed to separate different wavelength light in order of normal dispersion. (**f**) Schematic and (**g**) phase distribution of super-dispersion flat metalens M3 with anomalous dispersion. The distribution of focal points is opposite to M3. (**h**) Schematic and (**i**) phase distribution of super-dispersion flat metalens M4 with off-axis color separation. (**j**) Schematic and (**k**) phase distribution of super-dispersion flat metalens M5 with different off-axis colors separation. (Reproduced from [38] with permission).

In Table 2, it shows the different achromatic metasurface-based lens designs. The broadband achromatic lenses are pursued by researchers, with the great development of nano-fabrication technique and new materials applied in optical system, the performance of designed metasurface-based lenses will have a remarkable breakthrough.

Table 2. Comparison of metasurface-based lens in broadband achromatic optics design.

Reference (Year)	Material	NA	Wavelength
Khorasaninejad et al. (2017) [32]	TiO_2	0.2	490–550 nm
Wang et al. (2017) [35]	Au	0.324	1200–1680 nm
Li et al. (2017) [38]	Si	0.629	473, 532, and 632.8 nm

3.3. Metalens-Enabled Focus Scanning Devices

Miniaturized optical imaging and sensing systems commonly require focusing scanning or tuning mechanisms with ultra-compact size, lightweight, and fast scan speed. Micro-electro-mechanical system (MEMS) [39,40], electrowetting [41,42] and thermal tuning-based [43] methods have provided partial solutions. However, so far, these approaches are still relatively bulky and only provide a relatively slow tuning speed. Tunable focusing liquid crystal lenses [44–46] have been proposed for high tuning speed. However, due to the polarization dependence, the tuning speed is fully restricted. Metalenses have shown great potential for focus tuning applications. A reflective-mode metalens has been successfully integrated onto a high-speed MEMS scanner through the collaboration between Capasso's group at Harvard University and Lopez's group at the Argonne National Lab [47] for lateral beam scanning. Actuated by staggered comb-drives, the 2D MEMS scanner offers fast speed beam steering, high-resolution imaging, and high optical efficiency [48]. The fabricated metasurface lens was on a square substrate with side 0.8 mm and has a focal length of 5 mm with incident light 45° away from the surface, shown in Figure 8a. The metasurface lens was mounted at the center of the 2D MEMS scanner. To achieve biaxial scanning, the outer gimbal-frame rotates at a slow speed while the inner mirror scans with 9° tilting angle at high speed (~1 kHz). With an increasing voltage applied on the rotational axis, the MEMS scanner begins rotating until it reaches the maximum angles. The relationship between the applied voltage and the mechanical angle is shown in Figure 8b. To characterize the performance of the metasurface lens, the Finite Difference Time Domain (FDTD) method has been used to search for the electric field distribution (Figure 8c).

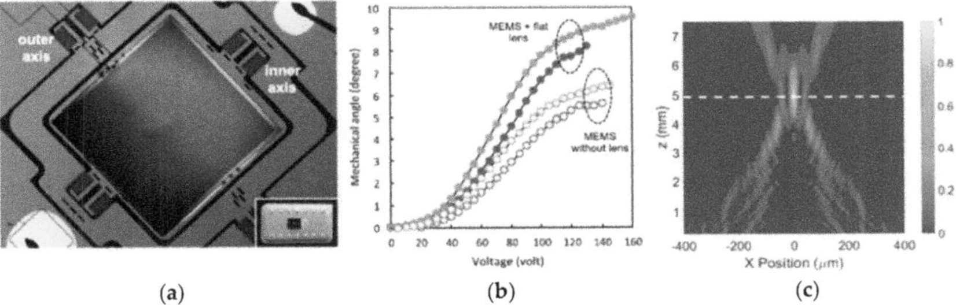

Figure 8. Metasurface-enabled MEMS (**a**) Optical microscope image of a MEMS scanner with a flat lens on top. (**b**) Angular displacement of the MEMS scanner with and without the metasurface-based lens. (**c**) Simulation: distribution of the intensity of the reflected beam in the xz-plane at $y = 0$. The maximum intensity occurred when $z = 5$ mm and focusing efficiency is 83%. (Reproduced from [47] with permission.)

Another metasurface-based MEMS device with tunable focus has been reported by Faraon's group [49] at Caltech. The new MEMS device was composed of two metalenses (a converging and diverging metasurface lenses), shown in Figure 9a. An axial focus scanning length can be achieved within a range of 565 to 629 μm. The first stationary metasurface lens was fabricated on a glass substrate and another moveable metasurface was fabricated on the SiN_x membrane. By applying a voltage potential, the distance between the two metalenses will change. By calculation, the various distance (Δx ~1 μm) results in a large tuning range (Δf ~ 36 μm). To tunable focal length, two series capacitors were placed on the substrates. In Figure 9b,c, the mechanical resonance of 2.6 kHz and 5.6 kHz was measured.

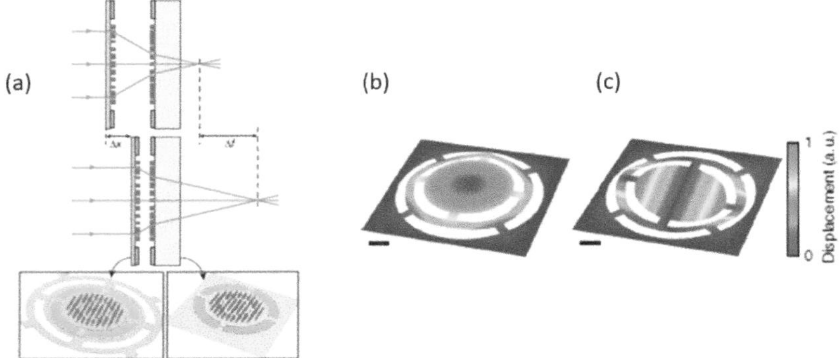

Figure 9. Metasurface-based MEMS device wit tunable focus. (**a**) Schematic illustration of the proposed tunable lens, comprised of a stationary lens on a substrate, and a moving lens on a membrane. The first (**b**) and second (**c**) mechanical resonances of the membrane at frequencies of 2.6 and 5.6 kHz, respectively. The scale bars are 100 μm. (Reproduced from [49] with permission.)

For testing the image qualities, the measurement step-up is shown in Figure 10a. When the object sets $p \sim 15$ mm away from the lens and no voltage applied, the image was out of focus. When we increased the voltage to 85 V, the image became clear. The same measurement was done for p equals 4 and 9.2 mm (Figure 10b).

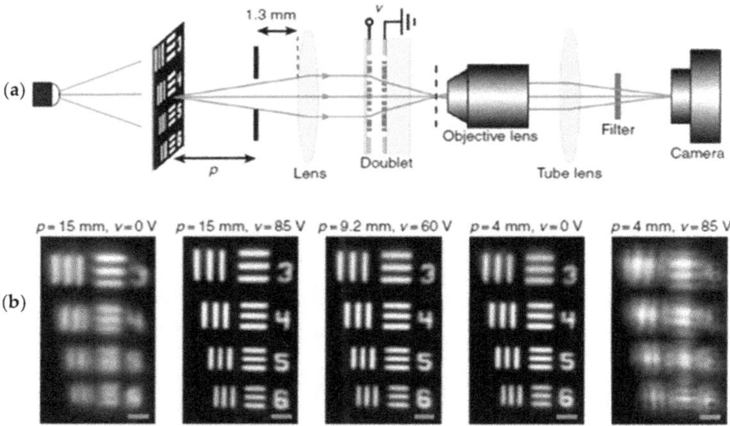

Figure 10. Imaging with the tunable doublet. (**a**) Schematic illustration of the imaging setup using a regular glass lens and the tunable doublet. The image formed by the doublet is magnified and re-imaged using a custom-built microscope with a ×55 magnification onto an image sensor. (**b**) Imaging results, showing the tuning of the imaging distance of the doublet and glass lens combination with applied voltage. By applying 85 V across the device, the imaging distance increases from 4 to 15 mm. The scale bars are 10 μm. (Reproduced from [49] with permission.)

3.4. Computational Optics Based on Metasurface

For bio-imaging applications, the most challenging problem is to image through scattering media (like human tissue specimens), because the passing light may have a complex speckle pattern. There are many existing methods to achieve high-resolution imaging, such as wavefront engineering [50], speckle correlations based on speckle correlations via non-invasive imaging through scattering layers [51,52], and a transmission matrix [53]. Recently, multiple computational imaging approaches have been demonstrated, including phase-space measurements [54,55], wavefront sensing with the Demon algorithm [56], and the speckle-correlation scattering matrix (SSM) [57,58]. For the conventional scattering media (CSM), lots of limitations still exist for practical applications in the real world, such as the stability of optical properties [59] and incomplete channel control in multiple scattering [60]. The trade-off between maximum scattering angle and memory-effect range [27] will ultimately cause significant defects from a practical point of view.

Recently, a few studies have been reported on the metasurface diffuser (MD), which can be used in wavefront control with a spatial light modulator [2,5,61–63]. The results show that the new imaging system will obtain a large FOV and high-resolution imaging quality. However, speckle patterns' computational imaging is less studied. Faraon's group [64] proposed a method combining MDs and SSM to replace the CSM complex field for 3D imaging. Researchers indicated that the MDs (Figure 11a) capture samples' amplitude and holographic imaging with numerical backpropagation. The nano-scatterers array (Figure 11b) provides 2π phase coverage and NA = 0.6.

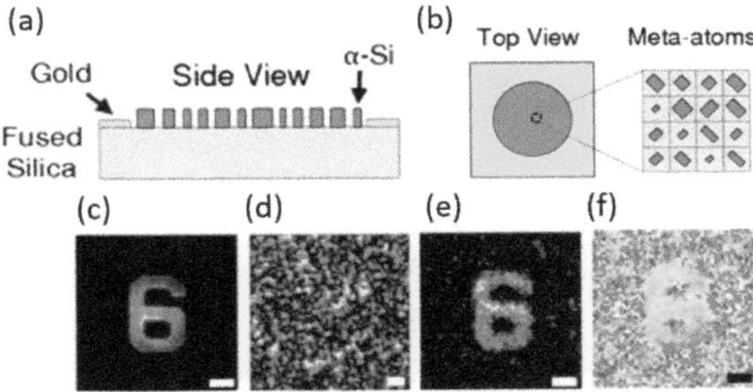

Figure 11. Metasurface-based computational imaging process. (**a**) Schematic illustration of the side view of the MD. (**b**) Schematics of a uniform array and unit cell of the metasurface, showing the parameter definitions. The transmission phase of the two orthogonal polarizations can be manipulated using the meta-atoms. (**c**) In-focus images of targets captured by a custom-built microscope. (**d**) The resulting speckle patterns of the samples after passing through the MD. (**e**) The retrieved object amplitudes. (**f**) Phases from the captured speckle patterns. The scale bars are 25 μm. (Reproduced from [64] with permission.)

To characterize the metasurface-based computational imaging system, the 1951 USAF resolution test target was used as an amplitude object (Figure 11c–f). As a result, CSM can be replaced by SSM and MD will overcome the trade-off limitations between efficiency and scattering angle.

4. A Metalens -Based Optical Imaging and Sensing System

Metalenses have been successfully integrated into state-of-the-art miniaturized bio-optical systems. Experimental results have validated the feasibility of the metalens for very broad applications in the future. The metalens holds promise for miniature optical imaging and sensing systems, such as optical coherence tomography (OCT), two-photon fluorescence microscopy, confocal microscopy, and spectrometers (Table 3). Details will be introduced in each session.

Table 3. Metalens-based optical imaging and sensing system.

Reference (Year)	Modality	Wavelength
Pahlevaninezhad et al. (2018) [78]	OCT	λ_{Ex} = 1310 nm
Arbabi et al. (2018) [91]	Two-photon Fluorescence	λ_{Ex} = 820 nm
Qiu et al. (2018) [93]	Confocal	λ_{Ex} = 660 nm
Zhu et al. (2017) [101]	Spectrometer	480–700 nm

4.1. Metalens-Based Endoscopic Optical Coherence Tomography (OCT)

Endoscopic optical coherence tomography (OCT) [65,66] has been developed as a non-invasive bio-imaging tool for early cancer detection, diagnosis, and cancer staging. Among the various bio-imaging modalities, confocal endomicroscopy can provide high resolution of tissue structures at the cellular level but lacks sufficient penetration depth. In contrast, OCT provides an adequate depth range (up to 1 mm) and FOV but the limited resolution (relatively lower lateral resolution). OCT was designed based on coherence interferometry [67,68], which obtains images from under-the-surface tissue structures [69,70]. New OCT tools with ultra-high image resolution have been demonstrated [71–77]. Shorter wavelengths with broader-bandwidth light sources potentially improve image resolution. However, the depth of penetration still does not meet the requirements for applications such as deep tissue scanning. Capasso's group and Suter's group at Harvard University have recently developed a new metalens-based endoscopic OCT [78] that can provide high-resolution imaging with extended depth.

As shown in Figure 12, the new OCT endoscope uses the metalens to replace the refractive lenses. However, the chromatic dispersion is a large defect in the system. Several techniques [79,80] have been reported to overcome or reduce the chromatic dispersion of metasurface. A phantom with a subwavelength gold line on the glass substrate was fabricated by electron-beam lithography (EBL). This new nano-optic endoscope was connected to the Fourier–domain OCT system [81,82]. The effective depth of focus will achieve 211 μm (tangential) and 315 μm (sagittal), larger than the achromatic lens, with the same NA that can provide (about 90 μm). Ex vivo imaging on freshly excised lung tissue has been demonstrated with the metalens-based nano-optic endoscope (Figure 12e), visualizing fine features in tissue, including epithelium, basement membrane, and cartilage.

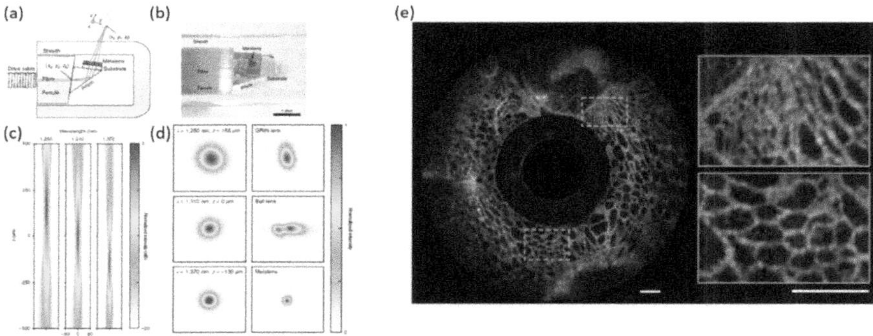

Figure 12. Metalens-based endoscopic optical coherence tomography (**a**) Schematic of the nano-optic endoscope. (**b**) Photographic image of the distal end of nano-optic endoscope. (**c**) Measured intensity distribution of the output beam of the nano-optic endoscope along the propagation direction in the *yz*-plane at λ = 1250, 1310 and 1370 nm. (**d**) (**left**) Focal spot profiles of the nano-optic endoscope at corresponding wavelengths. (**d**) (**right**) Focal spot profiles of a graded-index (GRIN) OCT catheter, a ball lens OCT catheter and the nano-optic endoscope at 1310 nm wavelength. (**e**) OCT images of fruit flesh (grape) obtained using a nano-optical endoscope. (Reproduced from [78] with permission.)

4.2. Metalens-Based Two-Photon Microscope

In biological study, high-resolution two-photon microscopy (nonlinear optics) has been widely used [83–86]. The development of a miniaturized two-photon microscope is still a challenge in the optics field. With the progress in dielectric metasurface research, researchers in Caltech have begun to construct a new two-photon microscope with tiny metalenses, because of the unique advantages such as lightweight, small structure, high efficiency, and controllable phase profiles. Commonly, metalenses are used for single-photon (linear optics) fluorescence images due to their limits with narrow effective wavelength. Metalenses usually operate at a single wavelength. However, for fluorescence microscopic imaging, metalenses will have excitation and emission wavelengths at different focal positions due to the chromatic dispersion [87–90]. This dispersion will reduce the collection efficiency of the system. A metalens-based two-photon microscope has been proposed by Arbabi et al. [91]. The designed system used double-wavelength metalenses (DW-ML) to replace the objective lenses. The DW-ML has the excitation and emission wavelength at the same focal distance and can acquire high-resolution fluorescence images compared to the traditional objectives. A schematic drawing of the metalens-based two-photon microscope is shown in Figure 13a. The incident high-peak-power pulsed laser passes through the DW-ML and is focused into the specimens. Collimated by the DW-ML, emitted fluorescent light (shorter wavelength) will be reflected by the dichroic mirror and collected by a detector. To realize the same focal length, the birefringent dichromic meta-atom method has been used [92]. The excitation wavelength is 820 nm near-infrared (NIR), while the collected emission light wavelength is 605 nm (Figure 13b). The nanofins are based on polycrystalline silicon (p-Si) due to its high index and low efficiency loss at the selected wavelength. Figure 13c shows the imaging characterization. Finally, the new metalens has been tested by replacing the traditional objectives in a table-top two-photon microscope. The acquired images of fluorophore-coated polyethylene microspheres are shown in Figure 13d,e. The quality of the DW-ML image is similar to that of conventional and bulky objectives. Although the DW-ML can provide a high-resolution image for a two-photon microscope, the number of collected photons is 15× lower and the required excited laser energy is 4.7× higher than a conventional objective.

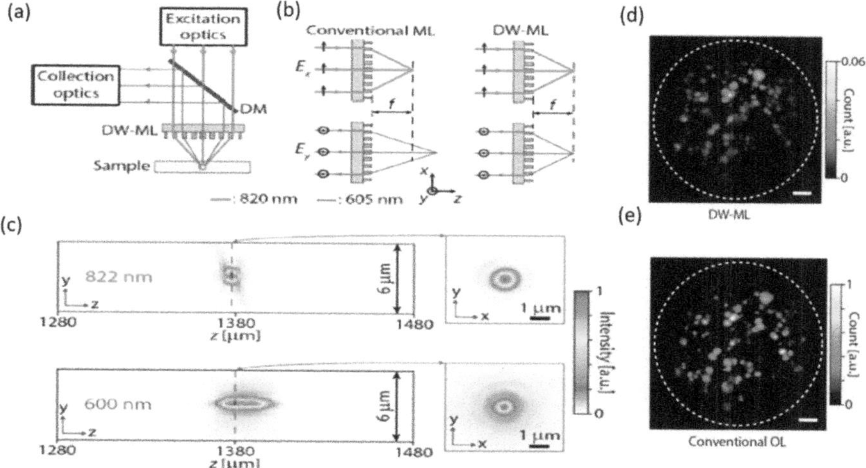

Figure 13. Metalens-based two-photon microscope system. (**a**) Schematic of a two-photon microscope employing a metasurface objective. (**b**) Schematic illustration of a conventional metasurface lens focusing light with different wavelengths to distinct focal lengths, and the DW-ML designed to focus 820 nm x-polarized light and 605 nm y-polarized light to the same focal distance of f. (**c**) (**Top**) Measured light intensity in the axial (**left**) and focal (**right**) planes for x-polarized 822 nm illumination. (**Bottom**) Same results as in the top figure for a y-polarized light source with a center wavelength of 600 nm and full width at half-maximum of approximately 10 nm. (**d**) Two-photon fluorescent microscope image captured by DW-ML. (**e**) captured using a conventional refractive objective, scale bars: 10 µm. (Reproduced from [91] with permission).

4.3. Metalens-Based Confocal Microscope System

The metalens has become an emerging technology for advanced miniature bio-medical optical imaging systems, including confocal microendoscopy for imaging hollow organs, which is one of the most promising imaging tools used in the clinic for "optical biopsy" applications. Qiu et al. [93] of Michigan State University recently demonstrated the feasibility of using a metalens for miniaturizing an optical fiber-based laser confocal microscope. The Huygens' metasurface-based metalens [94,95] has been made out of TiO_2 nano-disc shape resonators, which are polarization-independent (Figure 14a). With the support of crossed electric and magnetic dipole resonances, the incidence scattering direction will be cancelled, leaving only the forward propagation. By the ideal of designing the nanostructure geometry, the resonator arrays are able to achieve high transmission efficiency and 2π phase shift coverage (Figure 14b). The confocal microscope system consists of a photodetector, a data acquisition system, a CW incident laser wavelength of 660 nm, and Huygens' metalens. The system was characterized using the USAF resolution target. The measured focal length is 4 mm and the focused spot size is 4.29 μm (Figure 14c).

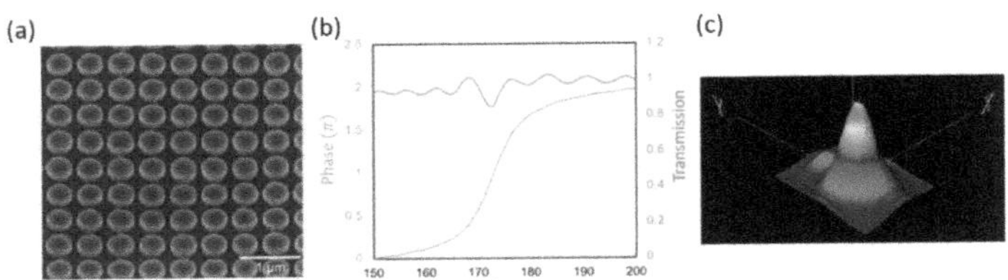

Figure 14. Metalens-based confocal microscope system. (**a**) TiO_2 nanoresonator array. (**b**) Phase and transmission vs. nanodisc radius by FDTD simulation. (**c**) a 3D intensity laser-focused spot profile. (Reproduced from [93] with permission.)

4.4. Metalens-Enabled Advanced Spectrometer

A miniaturized spectrometer system is another important and promising application for metalenses. Advanced spectrometers have been utilized for many applications, such as on-site environmental monitoring and in vivo disease diagnostics [96–98]. Currently, spectrometers are developed with conventional optics, such as non-polarizing beam splitters, wave plates, and polarizers. These spectrometer systems usually consist of focusing mirrors and grating turret [99]. Because of the limited grating dispersion, large spatial separation is hard to achieve within a short light propagation distance. Due to the limited properties of traditional optics, researchers are motivated to search for novel optics, such as metasurface-based lenses.

A new silicon-based off-axis metalens at near-infrared (NIR) wavelength has been demonstrated, which can provide higher spectral resolution than the conventional grating [100]. To improve the efficiency in the visible range, Capasso's group [101] also proposed an upgraded system, which operates in the visible wavelength range (Figure 15), adding an additional process—titanium oxide atomic layer deposition (ALD) [102]. The meta-spectrometer has several advantages: (1) focusing and dispersive elements will be placed in a planar structure; (2) it can surpass traditional blazed grating such as larger dispersions; (3) with different NA metasurface structures integrated on a single substrate, multiple spectral resolutions can be achieved.

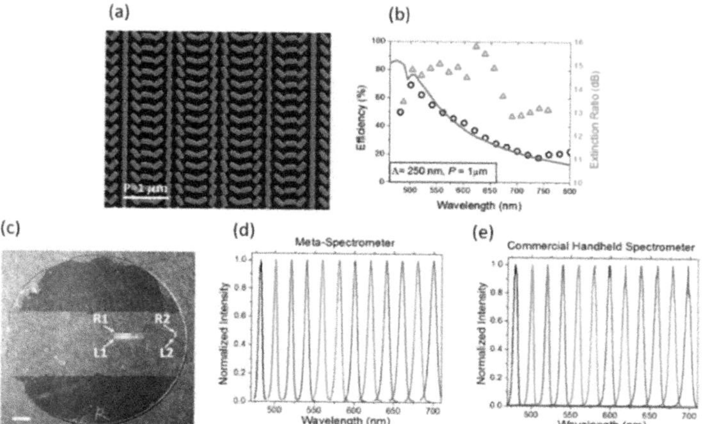

Figure 15. Ultra-compact visible chiral spectrometer with metalens. (**a**) Scanning electron microscope image of a fabricated off-axis metalens. (**b**) Measured (**black circles**) and simulated (**green line**) efficiencies of the +1 order metasurface grating under illumination with right-handed circularly polarized light. Extinction ratio (**red triangles**) of the meta-grating as a function of wavelength. (**c**) Photograph of a fabricated device with four separate metalenses labeled R1, R2, L1, and L2. (**d**) Measured spectra from a supercontinuum laser with 5 nm bandwidth using the metalens spectrometer. (**e**) Measured spectra from a commercial handheld spectrometer, the center wavelength varied from 480 nm to 700 nm. (Reproduced from [101] with permission.)

The challenges of refractive and diffractive limitations [103–106] still exist, such as curved focal plane and different wavelength aberrations, for the off-axis focusing metalenses. To resolve these, a newly developed aberration-corrected off-axis metalens [107], shown in Figure 16a, has a focal spot profile with a broad bandwidth, simultaneously engineering the phase and its higher order derivatives with respect to frequency (i.e., group delay (GD) and GD dispersion (GDD)). To achieve off-axis focusing, the phase profile of the metalens needed to satisfy Equation (3):

$$\varphi(x,y,\omega) = -\frac{\omega}{c}\left(\sqrt{\left(x-x_f\right)^2+y^2+\left(z-z_f\right)^2}-f(\omega)\right) \tag{3}$$

where x and y are spatial coordinates along the lens, and ω is the angular frequency of incident light. c is the light speed in vacuum. By using Taylor expanding on Equation (2) at designed angular frequency ω_d, the phase profile can be represented by Equation (4):

$$\varphi(x,y,\omega) = \varphi(x,y,\omega_d) + \frac{\partial\varphi}{\partial\omega}\cdot(\omega-\omega_d) + \frac{\partial^2\varphi}{2\partial\omega^2}\cdot(\omega-\omega_d)^2) \tag{4}$$

where the second term $\frac{\partial\varphi}{\partial w}$ of the partial derivative is the GD and the next higher term $\frac{\partial^2\varphi}{\partial\omega^2}$ represents GDD. The working distance is 4 cm and the spectral resolution can achieve 200 nm bandwidth in the visible range. To characterize the metalens performance, a traditional Berry phase lens was used as the comparison target. Each lens was illuminated with a collimated monochromatic laser at different wavelengths: 488, 532, 632, and 660 nm (Figure 16b).

The focal spots of the aberration-corrected metalens have the perfect single-peak signal for four different wavelengths. However, the focal spots for Berry phase lens become abnormal when the wavelength exceeds 532 nm. Another challenge for the aberration-corrected metalens is the dispersion.

A doublet-based metalens has been micro-machined to achieve strictly linear dispersion (Figure 17). The spectrometer can work for the helicity-polarized light.

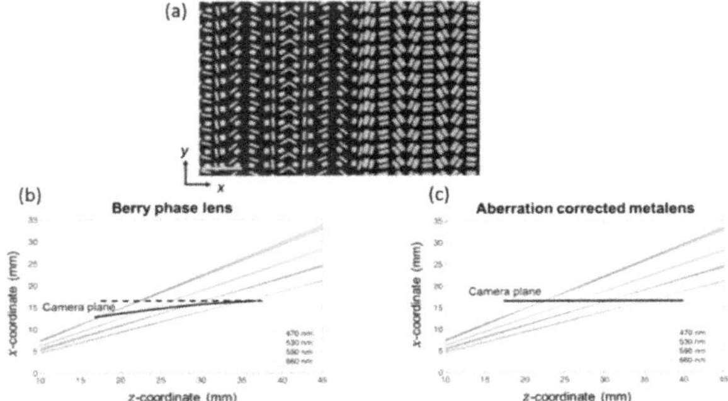

Figure 16. Compact aberration-corrected spectrometer. (**a**) Scanning electron micrographs of a fabricated aberration-corrected off-axis metalens. (**b**) A regular Berry-phase lens and (**c**) an aberration-corrected metalens. The metalenses designed with focal length $f = 40$ mm and focusing angle $a = 25$ degrees at wavelength $\lambda = 470$ nm. The focusing planes for each case are indicated by bold lines; the dashed line in the left is horizontal and meant as a reference to the curved focal plane. (Reproduced from [107] with permission.)

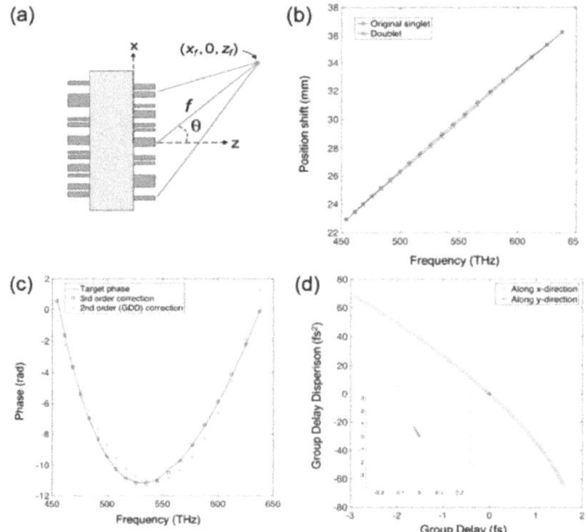

Figure 17. Compact aberration-corrected spectrometer characterizations. (**a**) Schematic of a doublet, comprising of a metasurface corrector and the original aberration-corrected off-axis metalens. The corrector serves to impart group delay (GD) and group delay dispersions (GDDs) such that the focal spot positions of the metalens are linear in frequency. (**b**) Plots of the focal spot positions of the singlet metalens (**blue**) and the doublet (**orange**) as a function of frequency. (**c**) Plot of the required phase as a function of frequency for an element at the edge of the metasurface corrector (**blue line**), and the results of second- and third-order polynomial fits (**orange crosses** and **black circles**, respectively). (**d**) GD and GDD values required for elements across the middle of the metasurface corrector, along x and y directions (**blue circles** and **orange crosses**). Inset: magnified view of the dispersion required for elements along y, which is minimal, due to the choice of the orthogonal camera plane. (Reproduced from [107] with permission.)

The compact spectrometer design successfully integrates a series of different wavelength band-pass filters on a single array of photodetectors [108,109]. However, such devices have system limitations in terms of resolution due to filter quality factors. Another design based on flat on-chip photonics can provide high spectral resolution [110–115], but the input on-chip coupling light losses and the reduced output are major challenges in many applications. To achieve a compact spectrometer with high resolution, Faraon's group [106] proposed a compact folded multi-metalenses system. With only 1 mm thickness, the new spectrometer can provide 1.2 nm resolution over a broad bandwidth range (~ 100 nm) in the NIR range. A schematic is shown in Figure 18a. Multiple metalenses are micro-machined on a transparent substrate to focus and disperse the incident light to different points. Ray-tracing simulation results are shown in Figure 18b, while the spot profiles are shown in Figure 18c. A small aberration was confirmed in the range of 760 nm to 860 nm. Using the diffraction-limited Airy radius and the focus displacement by the wavelength, the resolution was calculated in Figure 18d; the theoretical value is ~1.1 nm. To characterize the performance of the designed system, one-dimensional intensity profiles are shown in Figure 18e,f for transverse electric (TE) and transverse magnetic (TM).

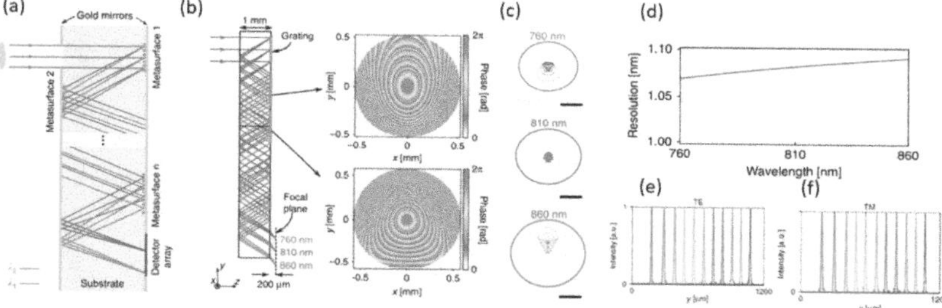

Figure 18. Compact folded metasurface spectrometer (**a**) The proposed scheme for a folded compact spectrometer. (**b**) Ray-tracing simulation results of the folded spectrometer, shown at three wavelengths in the center and two ends of the band. The system consists of a blazed grating that disperses light to different angles, followed by two metasurfaces optimized to focus light for various angles (corresponding to different input wavelengths). The grating has a period of 1 μm, and the optimized phase profiles for the two metasurfaces are shown on the right. (**c**) Simulated spot diagrams for three wavelengths: center and the two ends of the band. The scale bars are 5 μm. (**d**) Spectral resolution of the spectrometer, which is calculated from simulated Airy disk radii and the lateral displacement of the focus with wavelength. (**e**,**f**) One-dimensional focal spot profiles measured for several wavelengths in the bandwidth along the y-direction (as indicated in the inset) for TE and TM polarizations. The wavelengths start at 760 nm (blue curve) and increase at 10-nm steps up to 860 nm (red curve). (Reproduced from [106] with permission.)

5. Conclusions and Future Work

Metalenses has shown great potential for the development of miniaturized imaging and sensing systems. By replacing traditional lenses of bulky size, researchers have successfully demonstrated not only novel metalens designs but also advanced metalens-based optical systems with ultra-compact size. Many bio-optical applications will benefit from the remarkable advantages of the metalens, including ultrathin structure, large FOV, achromatic effects in a broad bandwidth, etc. Superior optical performance can be achieved by optimizing nanostructures on the metalens' surface. It has been proven that a metalens could potentially have a very high transmission or reflectivity efficiency. We believe that metalenses will have significant impacts on multiple imaging and sensing modalities, such as camera-based imaging systems, optical coherent tomography, two-photon, confocal, and spectrometer.

In a miniaturized optical system, metalenses could provide plenty of functions, including phase control, polarization, focus tuning, etc. Currently, most studies are still mainly focused on the fundamental properties of metalenses and their advanced microfabrication. In the future, metalens-based miniaturized bio-optical systems will attract more attention for broader biomedical applications, such as handheld, wearable, endoscopic, and implantable medical devices for quantitative healthcare.

Author Contributions: Writing—original draft preparation, B.L., W.P., Z.Q.; writing—review and editing, B.L., W.P., Z.Q.

Funding: This work is partially funded by the National Science Foundation (Grant number 1808436, to Z.Q.); the Department of Energy (Grant number 234402, to Z.Q.); the National Institutes of Health (NIH)/National Cancer Institute (NCI); the KMITL Research Fund; the National Research Council of Thailand; the Thailand Research Fund, Office of the Higher Education Commission of Thailand; the Newton Fund Researcher Links, British Council, UK; and the Fraunhofer-Bessel Research Award, Alexander von Humboldt Foundation, Germany.

Conflicts of Interest: The authors declare no conflict of interest.

References

1. Sun, S.; He, Q.; Xiao, S.; Xu, Q.; Li, X.; Zhou, L. Gradient-index meta-surfaces as a bridge linking propagating waves and surface waves. *Nat. Mater.* **2012**, *11*, 426. [CrossRef]
2. Yu, N.; Capasso, F. Flat optics with designer metasurfaces. *Nat. Mater.* **2014**, *13*, 139. [CrossRef]
3. Kildishev, A.V.; Boltasseva, A.; Shalaev, V.M. Planar Photonics with Metasurfaces. *Science* **2013**, *339*, 1232009. [CrossRef] [PubMed]
4. Glybovski, S.B.; Tretyakov, S.A.; Belov, P.A.; Kivshar, Y.S.; Simovski, C.R. Metasurfaces: From microwaves to visible. *Phys. Rep.* **2016**, *634*, 1–72. [CrossRef]
5. Jahani, S.; Jacob, Z. All-dielectric metamaterials. *Nat. Nanotechnol.* **2016**, *11*, 23. [CrossRef]
6. Chen, W.T.; Yang, K.-Y.; Wang, C.-M.; Huang, Y.-W.; Sun, G.; Chiang, I.D.; Liao, C.Y.; Hsu, W.-L.; Lin, H.T.; Sun, S.; et al. High-Efficiency Broadband Meta-Hologram with Polarization-Controlled Dual Images. *Nano Lett.* **2014**, *14*, 225–230. [CrossRef]
7. Knight, M.W.; Liu, L.; Wang, Y.; Brown, L.; Mukherjee, S.; King, N.S.; Everitt, H.O.; Nordlander, P.; Halas, N.J. Aluminum Plasmonic Nanoantennas. *Nano Lett.* **2012**, *12*, 6000–6004. [CrossRef]
8. Arbabi, A.; Horie, Y.; Bagheri, M.; Faraon, A. Dielectric metasurfaces for complete control of phase and polarization with subwavelength spatial resolution and high transmission. *Nat. Nanotechnol.* **2015**, *10*, 937. [CrossRef] [PubMed]
9. Lalanne, P.; Astilean, S.; Chavel, P.; Cambril, E.; Launois, H. Design and fabrication of blazed binary diffractive elements with sampling periods smaller than the structural cutoff. *J. Opt. Soc. Am. A* **1999**, *16*, 1143–1156. [CrossRef]
10. Vo, S.; Fattal, D.; Sorin, W.V.; Peng, Z.; Tran, T.; Fiorentino, M.; Beausoleil, R.G. Sub-Wavelength Grating Lenses with a Twist. *IEEE Photonics Technol. Lett.* **2014**, *26*, 1375–1378. [CrossRef]
11. Khorasaninejad, M.; Capasso, F. Metalenses: Versatile multifunctional photonic components. *Science* **2017**, *358*, eaam8100. [CrossRef]
12. Novotny, L.; van Hulst, N. Antennas for light. *Nat. Photonics* **2011**, *5*, 83. [CrossRef]
13. Yin, L.; Vlasko-Vlasov, V.K.; Pearson, J.; Hiller, J.M.; Hua, J.; Welp, U.; Brown, D.E.; Kimball, C.W. Subwavelength Focusing and Guiding of Surface Plasmons. *Nano Lett.* **2005**, *5*, 1399–1402. [CrossRef]
14. Liu, Z.; Steele, J.M.; Srituravanich, W.; Pikus, Y.; Sun, C.; Zhang, X. Focusing Surface Plasmons with a Plasmonic Lens. *Nano Lett.* **2005**, *5*, 1726–1729. [CrossRef]
15. Arbabi, A.; Horie, Y.; Ball, A.J.; Bagheri, M.; Faraon, A. Subwavelength-thick lenses with high numerical apertures and large efficiency based on high-contrast transmitarrays. *Nat. Commun.* **2015**, *6*, 7069. [CrossRef] [PubMed]
16. Khorasaninejad, M.; Zhu, A.Y.; Roques-Carmes, C.; Chen, W.T.; Oh, J.; Mishra, I.; Devlin, R.C.; Capasso, F. Polarization-Insensitive Metalenses at Visible Wavelengths. *Nano Lett.* **2016**, *16*, 7229–7234. [CrossRef]
17. Lianwei, C.; Yan, Z.; Mengxue, W.; Minghui, H. Remote-mode microsphere nano-imaging: New boundaries for optical microscopes. *OEA* **2018**, *1*, 170001. [CrossRef]
18. Feng, Q.; Pu, M.; Hu, C.; Luo, X. Engineering the dispersion of metamaterial surface for broadband infrared absorption. *Opt. Lett.* **2012**, *37*, 2133–2135. [CrossRef]
19. Pu, M.; Hu, C.; Wang, M.; Huang, C.; Zhao, Z.; Wang, C.; Feng, Q.; Luo, X. Design principles for infrared wide-angle perfect absorber based on plasmonic structure. *Opt. Express* **2011**, *19*, 17413–17420. [CrossRef]

20. Li, X.; Chen, L.; Li, Y.; Zhang, X.; Pu, M.; Zhao, Z.; Ma, X.; Wang, Y.; Hong, M.; Luo, X. Multicolor 3D meta-holography by broadband plasmonic modulation. *Sci. Adv.* **2016**, *2*, e1601102. [CrossRef]
21. Deng, Z.-L.; Zhang, S.; Wang, G.P. A facile grating approach towards broadband, wide-angle and high-efficiency holographic metasurfaces. *Nanoscale* **2016**, *8*, 1588–1594. [CrossRef]
22. Deng, Z.-L.; Li, G. Metasurface optical holography. *Mater. Today Phys.* **2017**, *3*, 16–32. [CrossRef]
23. Zhang, X.; Jin, J.; Pu, M.; Li, X.; Ma, X.; Gao, P.; Zhao, Z.; Wang, Y.; Wang, C.; Luo, X. Ultrahigh-capacity dynamic holographic displays via anisotropic nanoholes. *Nanoscale* **2017**, *9*, 1409–1415. [CrossRef]
24. Arbabi, A.; Arbabi, E.; Kamali, S.M.; Horie, Y.; Han, S.; Faraon, A. Miniature optical planar camera based on a wide-angle metasurface doublet corrected for monochromatic aberrations. *Nat. Commun.* **2016**, *7*, 13682. [CrossRef]
25. Groever, B.; Chen, W.T.; Capasso, F. Meta-Lens Doublet in the Visible Region. *Nano Lett.* **2017**, *17*, 4902–4907. [CrossRef]
26. Kingslake, R. *A History of the Photographic Lens*; Academic Press: Boston, MA, USA, 1989; p. xi. 334p.
27. Jang, M.; Horie, Y.; Shibukawa, A.; Brake, J.; Liu, Y.; Kamali, S.M.; Arbabi, A.; Ruan, H.; Faraon, A.; Yang, C. Wavefront shaping with disorder-engineered metasurfaces. *Nat. Photonics* **2018**, *12*, 84–90. [CrossRef]
28. Guo, Y.; Ma, X.; Pu, M.; Li, X.; Zhao, Z.; Luo, X. High-Efficiency and Wide-Angle Beam Steering Based on Catenary Optical Fields in Ultrathin Metalens. *Adv. Opt. Mater.* **2018**, *6*, 1800592. [CrossRef]
29. Xu, J.; Cua, M.; Zhou, E.H.; Horie, Y.; Faraon, A.; Yang, C. Wide-angular-range and high-resolution beam steering by a metasurface-coupled phased array. *Opt. Lett.* **2018**, *43*, 5255–5258. [CrossRef]
30. Pedrotti, F.L.; Pedrotti, L.M.; Pedrotti, L.S. *Introduction to Optics*; Cambridge University Press: Cambridge, UK, 2017.
31. Pu, M.; Li, X.; Ma, X.; Wang, Y.; Zhao, Z.; Wang, C.; Hu, C.; Gao, P.; Huang, C.; Ren, H.; et al. Catenary optics for achromatic generation of perfect optical angular momentum. *Sci. Adv.* **2015**, *1*, e1500396. [CrossRef]
32. Khorasaninejad, M.; Shi, Z.; Zhu, A.Y.; Chen, W.T.; Sanjeev, V.; Zaidi, A.; Capasso, F. Achromatic Metalens over 60 nm Bandwidth in the Visible and Metalens with Reverse Chromatic Dispersion. *Nano Lett.* **2017**, *17*, 1819–1824. [CrossRef]
33. Faklis, D.; Morris, G.M. Spectral properties of multiorder diffractive lenses. *Appl. Opt.* **1995**, *34*, 2462–2468. [CrossRef]
34. Hecht, E. *Optics*; Addison-Wesley: New York, NY, USA, 1997.
35. Wang, S.; Wu, P.C.; Su, V.-C.; Lai, Y.-C.; Hung Chu, C.; Chen, J.-W.; Lu, S.-H.; Chen, J.; Xu, B.; Kuan, C.-H.; et al. Broadband achromatic optical metasurface devices. *Nat. Commun.* **2017**, *8*, 187. [CrossRef] [PubMed]
36. Kurtzke, C. Suppression of fiber nonlinearities by appropriate dispersion management. *IEEE Photonics Technol. Lett.* **1993**, *5*, 1250–1253. [CrossRef]
37. Guo, Y.; Wang, Y.; Pu, M.; Zhao, Z.; Wu, X.; Ma, X.; Wang, C.; Yan, L.; Luo, X. Dispersion management of anisotropic metamirror for super-octave bandwidth polarization conversion. *Sci. Rep.* **2015**, *5*, 8434. [CrossRef]
38. Li, K.; Guo, Y.; Pu, M.; Li, X.; Ma, X.; Zhao, Z.; Luo, X. Dispersion controlling meta-lens at visible frequency. *Opt. Express* **2017**, *25*, 21419–21427. [CrossRef] [PubMed]
39. Lee, S.W.; Lee, S.S. Focal tunable liquid lens integrated with an electromagnetic actuator. *Appl. Phys. Lett.* **2007**, *90*, 121129. [CrossRef]
40. Shian, S.; Diebold, R.M.; Clarke, D.R. Tunable lenses using transparent dielectric elastomer actuators. *Opt. Express* **2013**, *21*, 8669–8676. [CrossRef]
41. Krogmann, F.; Mönch, W.; Zappe, H. A MEMS-based variable micro-lens system. *J. Opt. A Pure Appl. Opt.* **2006**, *8*, S330–S336. [CrossRef]
42. Li, L.; Wang, D.; Liu, C.; Wang, Q.-H. Zoom microscope objective using electrowetting lenses. *Opt. Express* **2016**, *24*, 2931–2940. [CrossRef]
43. Lee, S.; Tung, H.; Chen, W.; Fang, W. Thermal Actuated Solid Tunable Lens. *IEEE Photonics Technol. Lett.* **2006**, *18*, 2191–2193. [CrossRef]
44. Sato, S. Liquid-Crystal Lens-Cells with Variable Focal Length. *Jpn. J. Appl. Phys.* **1979**, *18*, 1679–1684. [CrossRef]
45. Ren, H.; Fan, Y.-H.; Gauza, S.; Wu, S.-T. Tunable-focus flat liquid crystal spherical lens. *Appl. Phys. Lett.* **2004**, *84*, 4789–4791. [CrossRef]

46. Pishnyak, O.; Sato, S.; Lavrentovich, O.D. Electrically tunable lens based on a dual-frequency nematic liquid crystal. *Appl. Opt.* **2006**, *45*, 4576–4582. [CrossRef] [PubMed]
47. Roy, T.; Zhang, S.; Jung, I.W.; Troccoli, M.; Capasso, F.; Lopez, D. Dynamic metasurface lens based on MEMS technology. *Apl Photonics* **2018**, *3*, 021302. [CrossRef]
48. Solgaard, O. *Photonic Microsystems: Micro and Nanotechnology Applied to Optical Devices and Systems*; Springer: New York, NY, USA, 2009; p. xvi. 631p.
49. Arbabi, E.; Arbabi, A.; Kamali, S.M.; Horie, Y.; Faraji-Dana, M.; Faraon, A. MEMS-tunable dielectric metasurface lens. *Nat. Commun.* **2018**, *9*, 812. [CrossRef] [PubMed]
50. Horstmeyer, R.; Ruan, H.; Yang, C. Guidestar-assisted wavefront-shaping methods for focusing light into biological tissue. *Nat. Photonics* **2015**, *9*, 563. [CrossRef] [PubMed]
51. Bertolotti, J.; van Putten, E.G.; Blum, C.; Lagendijk, A.; Vos, W.L.; Mosk, A.P. Non-invasive imaging through opaque scattering layers. *Nature* **2012**, *491*, 232. [CrossRef]
52. Katz, O.; Heidmann, P.; Fink, M.; Gigan, S. Non-invasive single-shot imaging through scattering layers and around corners via speckle correlations. *Nat. Photonics* **2014**, *8*, 784. [CrossRef]
53. Popoff, S.; Lerosey, G.; Fink, M.; Boccara, A.C.; Gigan, S. Image transmission through an opaque material. *Nat. Commun.* **2010**, *1*, 81. [CrossRef]
54. Takasaki, K.T.; Fleischer, J.W. Phase-space measurement for depth-resolved memory-effect imaging. *Opt. Express* **2014**, *22*, 31426–31433. [CrossRef]
55. Liu, H.-Y.; Jonas, E.; Tian, L.; Zhong, J.; Recht, B.; Waller, L. 3D imaging in volumetric scattering media using phase-space measurements. *Opt. Express* **2015**, *23*, 14461–14471. [CrossRef]
56. Berto, P.; Rigneault, H.; Guillon, M. Wavefront sensing with a thin diffuser. *Opt. Lett.* **2017**, *42*, 5117–5120. [CrossRef]
57. Lee, K.; Park, Y. Exploiting the speckle-correlation scattering matrix for a compact reference-free holographic image sensor. *Nat. Commun.* **2016**, *7*, 13359. [CrossRef]
58. Baek, Y.; Lee, K.; Park, Y. High-Resolution Holographic Microscopy Exploiting Speckle-Correlation Scattering Matrix. *Phys. Rev. Appl.* **2018**, *10*, 024053. [CrossRef]
59. Vellekoop, I.M.; Mosk, A.P. Focusing coherent light through opaque strongly scattering media. *Opt. Lett.* **2007**, *32*, 2309–2311. [CrossRef]
60. Goetschy, A.; Stone, A.D. Filtering Random Matrices: The Effect of Incomplete Channel Control in Multiple Scattering. *Phys. Rev. Lett.* **2013**, *111*, 063901. [CrossRef]
61. Ding, F.; Pors, A.; Bozhevolnyi, S.I. Gradient metasurfaces: A review of fundamentals and applications. *Rep. Prog. Phys.* **2017**, *81*, 026401. [CrossRef]
62. Hsiao, H.-H.; Chu, C.H.; Tsai, D.P. Fundamentals and Applications of Metasurfaces. *Small Methods* **2017**, *1*, 1600064. [CrossRef]
63. Genevet, P.; Capasso, F.; Aieta, F.; Khorasaninejad, M.; Devlin, R. Recent advances in planar optics: From plasmonic to dielectric metasurfaces. *Optica* **2017**, *4*, 139–152. [CrossRef]
64. Kwon, H.; Arbabi, E.; Kamali, S.M.; Faraji-Dana, M.; Faraon, A. Computational complex optical field imaging using a designed metasurface diffuser. *Optica* **2018**, *5*, 924–931. [CrossRef]
65. Tearney, G.J.; Boppart, S.A.; Bouma, B.E.; Brezinski, M.E.; Weissman, N.J.; Southern, J.F.; Fujimoto, J.G. Scanning single-mode fiber optic catheter–endoscope for optical coherence tomography. *Opt. Lett.* **1996**, *21*, 543–545. [CrossRef] [PubMed]
66. Tearney, G.J.; Brezinski, M.E.; Bouma, B.E.; Boppart, S.A.; Pitris, C.; Southern, J.F.; Fujimoto, J.G. In Vivo Endoscopic Optical Biopsy with Optical Coherence Tomography. *Science* **1997**, *276*, 2037. [CrossRef]
67. Youngquist, R.C.; Carr, S.; Davies, D.E.N. Optical coherence-domain reflectometry: A new optical evaluation technique. *Opt. Lett.* **1987**, *12*, 158–160. [CrossRef] [PubMed]
68. Huang, D.; Swanson, E.A.; Lin, C.P.; Schuman, J.S.; Stinson, W.G.; Chang, W.; Hee, M.R.; Flotte, T.; Gregory, K.; Puliafito, C.A.; et al. Optical coherence tomography. *Science* **1991**, *254*, 1178. [CrossRef] [PubMed]
69. Fujimoto, J.G.; De Silvestri, S.; Ippen, E.P.; Puliafito, C.A.; Margolis, R.; Oseroff, A. Femtosecond optical ranging in biological systems. *Opt. Lett.* **1986**, *11*, 150–152. [CrossRef] [PubMed]
70. Fercher, A.F.; Mengedoht, K.; Werner, W. Eye-length measurement by interferometry with partially coherent light. *Opt. Lett.* **1988**, *13*, 186–188. [CrossRef] [PubMed]

71. Liu, L.; Gardecki, J.A.; Nadkarni, S.K.; Toussaint, J.D.; Yagi, Y.; Bouma, B.E.; Tearney, G.J. Imaging the subcellular structure of human coronary atherosclerosis using micro–optical coherence tomography. *Nat. Med.* **2011**, *17*, 1010. [CrossRef] [PubMed]
72. Spöler, F.; Kray, S.; Grychtol, P.; Hermes, B.; Bornemann, J.; Först, M.; Kurz, H. Simultaneous dual-band ultra-high resolution optical coherence tomography. *Opt. Express* **2007**, *15*, 10832–10841. [CrossRef] [PubMed]
73. Cimalla, P.; Walther, J.; Mehner, M.; Cuevas, M.; Koch, E. Simultaneous dual-band optical coherence tomography in the spectral domain for high resolution in vivo imaging. *Opt. Express* **2009**, *17*, 19486–19500. [CrossRef]
74. Shu, X.; Beckmann, L.J.; Zhang, H.F. Visible-light optical coherence tomography: A review. *J. Biomed. Opt.* **2017**, *22*, 121707. [CrossRef]
75. Kim, J.; Xing, J.; Nam, H.S.; Song, J.W.; Kim, J.W.; Yoo, H. Endoscopic micro-optical coherence tomography with extended depth of focus using a binary phase spatial filter. *Opt. Lett.* **2017**, *42*, 379–382. [CrossRef]
76. Xi, J.; Zhang, A.; Liu, Z.; Liang, W.; Lin, L.Y.; Yu, S.; Li, X. Diffractive catheter for ultrahigh-resolution spectral-domain volumetric OCT imaging. *Opt. Lett.* **2014**, *39*, 2016–2019. [CrossRef]
77. Cui, D.; Chu, K.K.; Yin, B.; Ford, T.N.; Hyun, C.; Leung, H.M.; Gardecki, J.A.; Solomon, G.M.; Birket, S.E.; Liu, L.; et al. Flexible, high-resolution micro-optical coherence tomography endobronchial probe toward in vivo imaging of cilia. *Opt. Lett.* **2017**, *42*, 867–870. [CrossRef]
78. Pahlevaninezhad, H.; Khorasaninejad, M.; Huang, Y.-W.; Shi, Z.; Hariri, L.P.; Adams, D.C.; Ding, V.; Zhu, A.; Qiu, C.-W.; Capasso, F.; et al. Nano-optic endoscope for high-resolution optical coherence tomography in vivo. *Nat. Photonics* **2018**, *12*, 540–547. [CrossRef]
79. Fercher, A.F.; Hitzenberger, C.K.; Kamp, G.; El-Zaiat, S.Y. Measurement of intraocular distances by backscattering spectral interferometry. *Opt. Commun.* **1995**, *117*, 43–48. [CrossRef]
80. Haeusler, G.; Lindner, M.W. "Coherence radar" and "Spectral radar"—New tools for dermatological diagnosis. *J. Biomed. Opt.* **1998**, *3*, 21–31. [CrossRef]
81. Yun, S.H.; Tearney, G.J.; de Boer, J.F.; Bouma, B.E. Removing the depth-degeneracy in optical frequency domain imaging with frequency shifting. *Opt. Express* **2004**, *12*, 4822–4828. [CrossRef]
82. Yun, S.H.; Tearney, G.J.; Vakoc, B.J.; Shishkov, M.; Oh, W.Y.; Desjardins, A.E.; Suter, M.J.; Chan, R.C.; Evans, J.A.; Jang, I.-K.; et al. Comprehensive volumetric optical microscopy in vivo. *Nat. Med.* **2006**, *12*, 1429. [CrossRef]
83. Denk, W.; Strickler, J.H.; Webb, W.W. Two-photon laser scanning fluorescence microscopy. *Science* **1990**, *248*, 73. [CrossRef]
84. König, K. Multiphoton microscopy in life sciences. *J. Microsc.* **2000**, *200*, 83–104. [CrossRef]
85. Zipfel, W.R.; Williams, R.M.; Webb, W.W. Nonlinear magic: Multiphoton microscopy in the biosciences. *Nat. Biotechnol.* **2003**, *21*, 1369. [CrossRef]
86. Helmchen, F.; Denk, W. Deep tissue two-photon microscopy. *Nat. Methods* **2005**, *2*, 932. [CrossRef]
87. Miyamoto, K. The Phase Fresnel Lens. *J. Opt. Soc. Am.* **1961**, *51*, 17–20. [CrossRef]
88. Born, M.; Wolf, E. *Principles of Optics: Electromagnetic Theory of Propagation, Interference and Diffraction of Light*, 7th ed.; Cambridge University Press: Cambridge, UK; New York, NY, USA, 1999; p. xxxiii. 952p.
89. O'Shea, D.C. Society of Photo-optical Instrumentation Engineers. In *Diffractive Optics: Design, Fabrication, and Test*; SPIE Press: Bellingham, WA, USA, 2004; p. xii. 241p.
90. Arbabi, E.; Arbabi, A.; Kamali, S.M.; Horie, Y.; Faraon, A. Controlling the sign of chromatic dispersion in diffractive optics with dielectric metasurfaces. *Optica* **2017**, *4*, 625–632. [CrossRef]
91. Arbabi, E.; Li, J.; Hutchins, R.J.; Kamali, S.M.; Arbabi, A.; Horie, Y.; Van Dorpe, P.; Gradinaru, V.; Wagenaar, D.A.; Faraon, A. Two-Photon Microscopy with a Double-Wavelength Metasurface Objective Lens. *Nano Lett.* **2018**, *18*, 4943–4948. [CrossRef]
92. Arbabi, E.; Arbabi, A.; Kamali, S.M.; Horie, Y.; Faraon, A. High efficiency double-wavelength dielectric metasurface lenses with dichroic birefringent meta-atoms. *Opt. Express* **2016**, *24*, 18468–18477. [CrossRef]
93. Zhen, Q.; López, D.; Cai, H.; Piyawattanametha, W. Optical Fiber-Based Laser Confocal Microscope with a Metalens. In Proceedings of the 2018 International Conference on Optical MEMS and Nanophotonics (OMN), Lausanne, Switzerland, 29 July–2 August 2018; pp. 1–5.
94. Cai, H.; Czaplewski, D.A.; Stan, L.; López, D. High-efficiency, low-aspect-ratio planar lens based on Huygens resonators. In Proceedings of the 2017 International Conference on Optical MEMS and Nanophotonics (OMN), Santa Fe, NM, USA, 13–17 August 2017; pp. 1–2.

95. Zhang, L.; Ding, J.; Zheng, H.; An, S.; Lin, H.; Zheng, B.; Du, Q.; Yin, G.; Michon, J.; Zhang, Y.; et al. Ultra-thin high-efficiency mid-infrared transmissive Huygens meta-optics. *Nat. Commun.* **2018**, *9*, 1481. [CrossRef]
96. Zhu, H.; Isikman, S.O.; Mudanyali, O.; Greenbaum, A.; Ozcan, A. Optical imaging techniques for point-of-care diagnostics. *Lab Chip* **2013**, *13*, 51–67. [CrossRef]
97. Lunetta, R.S.; Elvidge, C. *Remote Sensing Change Detection: Environmental Monitoring Methods and Applications*; Taylor & Francis: London, UK, 1999; p. xviii. 318p.
98. Voller, A.; Bidwell, D.E.; Bartlett, A. Enzyme immunoassays in diagnostic medicine. Theory and practice. *Bull. World Health Organ.* **1976**, *53*, 55–65.
99. James, J.F.; Sternberg, R.S. *The Design of Optical Spectrometers*; Chapman & Hall: London, UK, 1969; p. xii. 239p.
100. Khorasaninejad, M.; Chen, W.T.; Oh, J.; Capasso, F. Super-Dispersive Off-Axis Meta-Lenses for Compact High Resolution Spectroscopy. *Nano Lett.* **2016**, *16*, 3732–3737. [CrossRef]
101. Zhu, A.Y.; Chen, W.-T.; Khorasaninejad, M.; Oh, J.; Zaidi, A.; Mishra, I.; Devlin, R.C.; Capasso, F. Ultra-compact visible chiral spectrometer with meta-lenses. *Apl Photonics* **2017**, *2*, 036103. [CrossRef]
102. Devlin, R.C.; Khorasaninejad, M.; Chen, W.T.; Oh, J.; Capasso, F. Broadband high-efficiency dielectric metasurfaces for the visible spectrum. *Proc. Natl. Acad. Sci. USA* **2016**, *113*, 10473. [CrossRef]
103. Zhang, C.; Cheng, G.; Edwards, P.; Zhou, M.-D.; Zheng, S.; Liu, Z. G-Fresnel smartphone spectrometer. *Lab Chip* **2016**, *16*, 246–250. [CrossRef]
104. Savage, N. Spectrometers. *Nat. Photonics* **2009**, *3*, 601. [CrossRef]
105. Gatkine, P.; Veilleux, S.; Hu, Y.; Bland-Hawthorn, J.; Dagenais, M. Arrayed waveguide grating spectrometers for astronomical applications: New results. *Opt. Express* **2017**, *25*, 17918–17935. [CrossRef]
106. Faraji-Dana, M.; Arbabi, E.; Arbabi, A.; Kamali, S.M.; Kwon, H.; Faraon, A. Compact folded metasurface spectrometer. *Nat. Commun.* **2018**, *9*, 4196. [CrossRef]
107. Zhu, A.Y.; Chen, W.T.; Sisler, J.; Yousef, K.M.A.; Lee, E.; Huang, Y.-W.; Qiu, C.-W.; Capasso, F. Compact Aberration-Corrected Spectrometers in the Visible Using Dispersion-Tailored Metasurfaces. *Adv. Opt. Mater.* **2018**, *0*, 1801144. [CrossRef]
108. Wang, S.-W.; Xia, C.; Chen, X.; Lu, W.; Li, M.; Wang, H.; Zheng, W.; Zhang, T. Concept of a high-resolution miniature spectrometer using an integrated filter array. *Opt. Lett.* **2007**, *32*, 632–634. [CrossRef]
109. Horie, Y.; Arbabi, A.; Arbabi, E.; Kamali, S.M.; Faraon, A. Wide bandwidth and high resolution planar filter array based on DBR-metasurface-DBR structures. *Opt. Express* **2016**, *24*, 11677–11682. [CrossRef]
110. Momeni, B.; Hosseini, E.S.; Askari, M.; Soltani, M.; Adibi, A. Integrated photonic crystal spectrometers for sensing applications. *Opt. Commun.* **2009**, *282*, 3168–3171. [CrossRef]
111. Pervez, N.K.; Cheng, W.; Jia, Z.; Cox, M.P.; Edrees, H.M.; Kymissis, I. Photonic crystal spectrometer. *Opt. Express* **2010**, *18*, 8277–8285. [CrossRef]
112. Xia, Z.; Eftekhar, A.A.; Soltani, M.; Momeni, B.; Li, Q.; Chamanzar, M.; Yegnanarayanan, S.; Adibi, A. High resolution on-chip spectroscopy based on miniaturized microdonut resonators. *Opt. Express* **2011**, *19*, 12356–12364. [CrossRef]
113. Redding, B.; Liew, S.F.; Sarma, R.; Cao, H. Compact spectrometer based on a disordered photonic chip. *Nat. Photonics* **2013**, *7*, 746. [CrossRef]
114. Nitkowski, A.; Chen, L.; Lipson, M. Cavity-enhanced on-chip absorption spectroscopy using microring resonators. *Opt. Express* **2008**, *16*, 11930–11936. [CrossRef] [PubMed]
115. Gan, X.; Pervez, N.; Kymissis, I.; Hatami, F.; Englund, D. A high-resolution spectrometer based on a compact planar two dimensional photonic crystal cavity array. *Appl. Phys. Lett.* **2012**, *100*, 231104. [CrossRef]

MDPI
St. Alban-Anlage 66
4052 Basel
Switzerland
Tel. +41 61 683 77 34
Fax +41 61 302 89 18
www.mdpi.com

Micromachines Editorial Office
E-mail: micromachines@mdpi.com
www.mdpi.com/journal/micromachines